FINANCING ENERGY EFFICIENCY

FINANCING ENERGY EFFICIENCY

Lessons from
Brazil, China, India, and Beyond

Robert P. Taylor
Chandrasekar Govindarajalu
Jeremy Levin
Anke S. Meyer
William A. Ward

©2008 The International Bank for Reconstruction and Development / The World Bank
1818 H Street NW
Washington DC 20433
Telephone: 202-473-1000
Internet: www.worldbank.org
E-mail: feedback@worldbank.org

All rights reserved

1 2 3 4 5 11 10 09 08

This volume is a product of the staff of the International Bank for Reconstruction and Development / The World Bank. The findings, interpretations, and conclusions expressed in this volume do not necessarily reflect the views of the Executive Directors of The World Bank or the governments they represent.

The World Bank does not guarantee the accuracy of the data included in this work. The boundaries, colors, denominations, and other information shown on any map in this work do not imply any judgement on the part of The World Bank concerning the legal status of any territory or the endorsement or acceptance of such boundaries.

Rights and Permissions

The material in this publication is copyrighted. Copying and/or transmitting portions or all of this work without permission may be a violation of applicable law. The International Bank for Reconstruction and Development / The World Bank encourages dissemination of its work and will normally grant permission to reproduce portions of the work promptly.

For permission to photocopy or reprint any part of this work, please send a request with complete information to the Copyright Clearance Center Inc., 222 Rosewood Drive, Danvers, MA 01923, USA; telephone: 978-750-8400; fax: 978-750-4470; Internet: www.copyright.com.

All other queries on rights and licenses, including subsidiary rights, should be addressed to the Office of the Publisher, The World Bank, 1818 H Street NW, Washington, DC 20433, USA; fax: 202-522-2422; e-mail: pubrights@worldbank.org.

ISBN-13: 978-0-8213-7304-0
eISBN-13: 978-0-8213-7305-7
DOI: 10.1596/978-0-8213-7304-0

Library of Congress Cataloging-in-Publication Data

Financing energy efficiency : lessons from Brazil, China, India, and beyond / by Robert P. Taylor ... [et al.].
  p. cm.
Includes bibliographical references and index.
ISBN 978-0-8213-7304-0—ISBN 978-0-8213-7305-7 (electronic)
  1. Industries—Energy consumption. 2. Industries—Energy conservation—Finance. 3. Energy policy. I. Taylor, Robert P. (Robert Prescott), 1955–II. World Bank.

HD9502.A2F565 2007
333.79′17—dc22
                                    2007039296

CONTENTS

Foreword   *xi*
Acknowledgments   *xiii*
Acronyms and Abbreviations   *xv*

Overview   1
Energy Efficiency Financing and the Three Country
    Energy Efficiency Project   1
The Need for Energy Efficiency Investment
    Financing Interventions   3
Delivery of Energy Efficiency Financing Is an
    Institutional Development Issue   7
Delivering Investment Project Designs and
    Technical Appraisals   8
Delivering Financing   10
Making Integrated Mechanisms Work   13
Moving Ahead   17

PART I. LESSONS FROM ENERGY EFFICIENCY FINANCING OPERATIONS IN CHINA, INDIA, AND BRAZIL

Chapter 1. Introduction   **23**
Energy and Growth   24

Why Energy Efficiency Is Important	27
The Three Country Energy Efficiency Project	30
Roadmap for this Book	31

Chapter 2. Summary of the Energy Efficiency Terrain   **35**
Reclassifying the Energy Efficiency Terrain	36
Improving Energy Efficiency in New Facilities	36
Improving Energy Efficiency in Existing Facilities	38
Conclusions	42
The Focus of this Book: Standard Energy Efficiency Projects	45

Chapter 3. Orgins and Persistence of Energy Inefficiency   **49**
The Importance of Institutional Environments for Energy Efficiency Investment	51
Additional Challenges Posed by Institutional Frameworks in Developing Market Economies	56
Production versus Efficiency Investment in Rapidly Growing Economies	59

Chapter 4. Models for Delivering Energy Efficiency Investments   **63**
A General Model for Successful Delivery Programs for Energy Efficiency Investment	65
Examples of Delivery Mechanisms for Energy Efficiency Investments	71

Chapter 5. Identifying and Developing Energy Efficiency Investment Projects   **79**
Market Selection and Outreach	80
Project Development: Identifying, Clarifying, and Allocating Risk	84
Institutional Capacities for Energy Efficiency Project Development	88
Making Choices about Outsourcing	92
Some Options to Minimize Transaction Costs	96

Chapter 6. Delivery of Financing   101
What Needs to be Delivered?   101
The Institutional Environment for Financing:
    The Different Worlds of Bankers and Energy
    Efficiency Project Promoters   102
Dealing with Banking Systems in Transition
    or under Development   105
Institutional Options for Delivery of Financing   107
Dealing with Repayment Issues in Energy Efficiency
    Projects and New Financing Product Development   110

Chapter 7. Making Investment Delivery Mechanisms Work   117
Basic Principles   118
Energy Efficiency Lending through Local Financial
    Institutions   120
Energy Service Companies (ESCOs)   131
Energy Utility Demand-Side Management (DSM)   138

Chapter 8. Conclusions and Recommendations   141
Suggestions for Each of the Three Countries   142
Suggested Roles for International Financial
    Institutions   149
Operational Suggestions on Development of
    New Projects   152

PART II. ENERGY EFFICIENCY FINANCE CASE STUDIES

Introduction to Part II   157
1. China ESCO Loan Guarantee Program   162
2. Hungary Energy Efficiency
    Guarantee Fund   170
3. Romania Energy Efficiency Fund   181
4. IREDA Energy Efficiency Loan Fund   189
5. Energy Efficiency Cluster Lending for
    SMEs by Indian Banks   194

6. Lithuania Energy Efficiency and
    Housing Pilot Project 205
 7. China's Full-Service ESCOs 213
 8. ESCO Development in the United States
    and Canada 224
 9. Brazil Public Benefit Wire-Charge Mechanism 235
10. Sri Lanka DSM: Using the Utility
    Bill as a Loan Repayment Mechanism 243
11. Dongying Shengdong EMC Waste Gas Power Projects 246
12. Iqara Energy Services in Brazil 250
13. India Capacitor Leasing 257

BIBLIOGRAPHY 261

**APPENDIX. GLOSSARY OF SELECTED TERMS IN
NEW INSTITUTIONAL ECONOMICS (NIE) THAT
RELATE TO ENERGY EFFICIENCY FINANCE** 265

INDEX 273

Boxes
  1.1 Energy Efficiency Investments
      Are Very Cost-Effective 29
  2.1 Why Distinguish Between "Restructuring
      Projects" and "Standard Energy
      Efficiency Projects?" 39
  4.1 Generalized Model for Developing
      New Energy Efficiency Investment
      Delivery Mechanisms in
      Developing Countries 68
  7.1 One Example of a Failed Project 121
CS9.1 Summary of Some Ideas for
      Reform of Brazil's Wire-Charge 241

Figures
  1.1 Growth of Developing
      Countries' Energy Demand 25

1.2	Energy-Related CO_2 Emissions Growth to 2030	27
7.1	Shared Savings EPC Model	132
7.2	Guaranteed Savings EPC Model	133
CS1.1	Structural Overview of the EMC Loan Guarantee Program	164
CS2.1	Hungary Energy Efficiency Co-financing Program Institutional Arrangements	171
CS2.2	HEECP Results, 1997–2006	174
CS3.1	FREE Institutional Arrangements and Funds Flow	184
CS4.1	IREDA Institutional Arrangements	190
CS5.1	Cluster Lending Approach Adopted in India	197
CS6.1	Lithuania Energy Efficiency Project Institutional Arrangements	206
CS7.1	China's Full-Service Shared Savings ESCO Model	214
CS7.2	Types of Projects Implemented 1997–2006 by Three Chinese ESCOs	219
CS8.1	Shared Savings Contracting Model	227
CS8.2	Guaranteed Savings Contracting Model	227
CS9.1	Breakdown of Brazilian Utilities' Energy Efficiency Investments by Sector (1998–2003)	238
CS9.2	Total Utilities' Energy Efficiency Investments by Sector (1998–2003)	238
CS10.1	Sri Lanka DSM Project: Institutional Arrangements	244
CS11.1	Dongying Shengdong EMC Ownership and Business Arrangements	247
CS12.1	Iqara's Business Model	252
CS13.1	Institutional Arrangements in a Capacitor Leasing Project in India	258

Tables

1.1	World Primary Energy Demand by Region, Mtoe (Reference Scenario)	25

2.1	Energy Efficiency Interventions by Economic Sector	37
2.2	Typical Policy and Regulatory Tools to Promote Energy Efficiency in New Facilities	38
3.1	Contract Enforcement: Brazil, China, and India Compared to Canada and the United States	57
CS2.1	Evolution of HEECP Parameters, 1997–2006	173
CS3.1	Romania Financial Market Conditions	182
CS3.2	Free Project Results	183
CS3.3	Summary of Advantages and Disadvantages of FREE	187
CS4.1	Pros and Cons of Supporting Energy Efficiency Investments through a Parastatal Entity	192
CS6.1	Lithuania Energy Efficiency Project Results	208
CS7.1	China ESCO EPC Project Investment 2005/2006	217
CS9.1	Allocation of Wire-Charge Uses in Brazil, 1998–2007	236
CS9.2	Total Investment in Regulated Utility Energy Efficiency Programs in Brazil	237
CS9.3	ESCO Contracts with Brazilian Utilities in Energy Efficiency Regulated Programs	239
CS12.1	Iqara Projects	253

FOREWORD

Greater energy efficiency is key for shifting country development paths toward lower-carbon economic growth. Especially in developing countries and transition economies, vast potential for energy savings opportunities remain unrealized even though current financial returns are strong. *Financing Energy Efficiency: Lessons from Brazil, China, India, and Beyond* examines the nature of this dilemma and how it may be overcome in practical and operational terms.

Tapping more aggressively into the wealth of available, financially attractive energy-saving renovation projects requires mechanisms to develop and deliver large numbers of relatively small projects scattered among hundreds of thousands of industries and building complexes. The investment opportunities result in operating-cost savings, as opposed to new production, and are technically and logistically diverse. As such, they often do not compete well with other opportunities for using up-front capital, such as capacity expansion or penetrating new markets. If left unaddressed, problems of prevailing high transaction costs, perceptions of uncertain risks, and unmet needs for financial intermediation or technical expertise mean that much of the potential for energy savings will remain unimplemented. Institutional innovation is required to address these problems and put in place efficient ways of identifying, packaging, and delivering bundles of energy saving projects.

This book reviews the reasons for the success or failure of a range of recent energy efficiency programs in developing countries and economies in transition. It also draws heavily on an intensive program to exchange ideas and operational lessons learned in energy efficiency projects in Brazil, China, and India, undertaken during 2002–06 with funding from the United Nations Foundation. The book attempts to synthesize lessons learned from the many practical experiences shared by scores of seasoned professionals working on energy efficiency in these three large developing countries and the thoughts of these practitioners on how to overcome the obstacles faced.

The book goes beyond those experiences to review lessons learned in various energy efficiency programs in recent years, especially by the World Bank. Part II of the book presents 13 case studies of specific energy efficiency investment delivery mechanisms implemented in eight different countries. One of the recurring themes of the book is that effective delivery of energy efficiency investments is essentially an institutional development challenge. As such, attempted solutions must fit within prevailing local economic institutional contexts, which vary dramatically. Where initiatives have been most successful, they have been built following careful, in-country diagnostic work, with parallel attention to both financial intermediation and technical support requirements and with flexibility to make many adjustments along the way.

Improving energy efficiency has become an increasingly urgent imperative across the world. It is our hope that this book can contribute to the further generation of new ideas and approaches on how to scale up energy efficiency investment.

Jamal Saghir
Director, Energy, Transport, and Water
Chair, Energy Sector Board
Sustainable Development Network
The World Bank

ACKNOWLEDGMENTS

This book was prepared by a World Bank team, led by Robert P. Taylor, consisting of Chandrasekar Govindarajalu, Jeremy Levin, Anke S. Meyer, and William A. Ward (Clemson University), with major contributions by Alan Poole, Shen Longhai, Alex Potter, Kapil Thukral, and Chuck Guinn.

The book draws heavily on the results of the multiyear, global technical assistance project called "Developing Financial Intermediation Mechanisms for Energy Efficiency Projects in Brazil, China, and India," (also known as the Three Country Energy Efficiency Project or the 3CEE project). The project was funded by the Energy Sector Management Assistance Program (ESMAP) and the United Nations Foundation (UNF), and was implemented jointly by the World Bank and the United Nations Environment Program (UNEP) Risoe Centre on Energy, Climate and Sustainable Development (URC) team of Jyoti P. Painuly, Juan Zak, and Myung K. Lee. Mark Radka at UNEP provided key support in project implementation.

For almost six years, a large number of officials, bankers, and energy efficiency specialists in the three countries have searched together with the study team for the means to overcome the various market barriers to major increases in domestic financing of energy efficiency investments. We would especially like to thank the individuals who participated from the country secretariats of the Three Country Energy

Efficiency Project: Maria Cecilia Amaral, Ricardo David, Newton Figueiredo, Nelson Pedrozo, and Nelson Albuquerque in Brazil; Shen Long Hai and Zhu Xingshan in China; and Debashish Majumdar, Debjani Bhatia, and Koshy Cherail in India. Full listings of the country participants are available in the annexes of the full Project Country Reports cited in the bibliography.

Sandy Selman, Tom Stoner, Pierre Langlois, and Shirley Hansen provided their expertise and guidance in implementation of several international cross-exchange workshops and in-country seminars.

The financial support of ESMAP and the UNF is gratefully acknowledged. Ms. Dominique Lallement was an early and strong supporter of this type of south-south exchange. Support from William Kennedy at the United Nations Fund for International Partnerships is also gratefully acknowledged.

The development of the project concept and this book benefited from advice and input provided by Peter Johansen and John MacLean (peer reviewers) and many others, inside and outside the World Bank. While contributors are too numerous to name, the project team remains indebted to those many experts and energy efficiency practitioners who provided insights from their experience.

Finally, the project could not have been implemented without continuing support from Carla Sarmiento, Norma Leon, Teri Velilla, Cristina Hernandez, Valeriya Goffe, Grace Aguilar, and Priya Chopra.

ACRONYMS AND ABBREVIATIONS

3CEE	Three Country Energy Efficiency
ABESCO	Brazilian Association of Energy Service Companies
AEC	Ahmedabad Electric Company
ANEEL	Brazilian National Power Regulatory Agency
BEE	Bureau of Energy Efficiency (India)
BgEEF	Bulgarian Energy Efficiency Fund
BNDES	Brazilian National Bank for Economic and Social Development
CEB	Ceylon Electricity Board
CECIC	China Energy Conservation Investment Corporation
CEEF	Commercializing Energy Efficiency Finance
CFL	Compact fluorescent lamp
CO_2	Carbon Dioxide
DFI	Development Finance Institution
DSM	Demand-side management
EBRD	European Bank for Reconstruction and Development
EE	Energy efficiency
EES	Enron Energy Services
EMC	Energy Management Company (Chinese term for ESCO)
EMCA	Chinese Energy Management Company Association
EPC	Energy Performance Contract

ESCO	Energy Service Company
ESMAP	Energy Sector Management Assistance Program
ESPC	Energy Savings Performance Contracts
FBI	Federal Buildings initiative
FEMP	Federal Energy Management Program
FI	Financial institution
FREE	Romanian Energy Efficiency Fund
GDP	Gross domestic product
GEF	Global Environment Facility
GFA	Guarantee Facility Agreement
GHG	Greenhouse gas
HEECP	Hungary Energy Efficiency Co-Financing Program
HOA	Homeowners Association
HUDF	Housing and Urban Development Foundation (Lithuania)
HVAC	Heating, ventilation and air conditioning
I&G	China National Investment and Guaranty Company
IBRD	International Bank for Reconstruction and Development
ICPEEB	Indian Council for Promotion of Energy Efficiency Businesses
IDA	International Development Association
IEA	International Energy Agency
IES	Iqara Energy Services
IFC	International Finance Corporation
IFI	International Financial Institution
IPO	Initial Public Offering
IREDA	Indian Renewable Energy Development Agency
IRR	Internal Rate of Return
MOU	Memorandum of Understanding
M&E	Monitoring and Evaluation
MDB	Multilateral Development Bank
MNRE	Ministry of New and Renewable Energy
NASEO	National Association of State Energy Officials
NBFI	Nonbank financial institution
NGO	Nongovernmental Organization

NIE	New Institutional Economics
OECD	Organisation for Economic Co-operation and Development
O&M	Operation and Maintenance
OP 5	GEF Operational Program Number 5
PROCEL	National Electricity Conservation Program, Brazil
SBI	State Bank of India
SIDBI	Small Industries Development Bank of India
SME	Small and medium enterprise
SSI	Small-scale industry
SSL	Saha Sprague Limited
TA	Technical assistance
TBSE	Technology Bureau for Small Enterprises
TGA	Transaction Guarantee Agreement
UNEP	United Nations Environment Program
UNF	United Nations Foundation
URC	Risoe Center of the United Nations Environment Program
USAID	U.S. Agency for International Development
WEO	World Energy Outlook

Weights, Measures, and Conversion Factors

GW	Gigawatt = 10^9
kWh	Kilowatt-hour
MW	Megawatt = 10^6 watts
boe	Barrel of Oil Equivalent
btoe	Billion Tons of Oil Equivalent
mtoe	Million Tons of Oil Equivalent
tce	Tons of Coal Equivalent
toe	Tons of Oil Equivalent

Exchange Rates

As of June 30, 2007

	Brazil (Real)	China (RMB)	India (Rupee)
US$1 =	1.929	7.625	40.735

OVERVIEW

ENERGY EFFICIENCY FINANCING AND THE THREE COUNTRY ENERGY EFFICIENCY PROJECT

New or improved programs to better capture the enormous potential for energy savings in existing industries and buildings in the developing world have important roles to play for the environment and for economic development.

Many thousands of energy efficiency projects with strong financial rates of return remain unimplemented in the world at large, but especially in developing countries and emerging markets. The essential issue blocking the realization of the potential energy savings is the underdeveloped state of energy efficiency investment delivery mechanisms, adapted to be able to work well in national and local economic environments. Traditional investment delivery mechanisms operated by local banks and other financing organizations often have played useful roles in the energy efficiency business, but still only a fraction of the potential has been tapped. Renewed and strong efforts are required to develop financing programs that can combine effective technical project development with financial products appropriate for dispersed investments, with benefits focused on operating cost savings. Many programs in recent years have aimed to develop such delivery mechanisms. Some have succeeded and some have failed. Given the urgent need to mobilize large levels of investments in energy efficiency to help meet future energy requirements, this book evaluates the experience of past efforts, attempts to summarize lessons learned, and provides suggestions on how these lessons may be applied in the future. The book concentrates on Brazil, China, and India, but also includes reviews of selected experiences in other countries.

This book draws extensively on the experiences of the UNEP-World Bank multiyear, technical assistance effort, "Developing Financial Intermediation Mechanisms for Energy Efficiency Projects in Brazil, China, and India" (also known as the Three Country Energy Efficiency Project), funded by the UNF and ESMAP. The purpose of this project was to generate new ideas and approaches for developing energy efficiency financing schemes, which then could be tried out by local institutions, with support from the World Bank and other international agencies and donors where necessary. Core groups of representatives from both the financing and energy efficiency development communities in each of the three countries implemented project activities. Energy efficiency and banking industry practitioners from each country also met in four formal cross-exchange workshops, and various informal meetings, to exchange lessons learned and ideas. This book attempts to synthesize the considerable practical knowledge generated from the project, which is applicable across countries, together with additional knowledge from other World Bank Group and donor efforts in other countries.

Following an introduction in chapter 1, chapter 2 summarizes the overall energy efficiency terrain and identifies the opportunities at which the recommendations of this book are directed. Chapter 3 explores the origins and persistence of energy inefficiency. Chapter 4 provides a framework for thinking about the basic organizational and institutional challenges and the basic types of energy efficiency investment mechanisms, and begins the discussion of the various mechanisms that have been used to meet these challenges. Chapter 5 discusses the need that all such investment mechanisms have for market identification and outreach, project development, and technical assessment of energy efficiency projects. Options are laid out for developing and incorporating needed local technical capacity within investment delivery mechanisms. Chapter 6 deals with arranging the financing flows that all investment mechanisms require, the issues involved, and the available options for financing. Chapter 7 summarizes experience with the development and operation of a range of energy efficiency investment mechanisms, and some of the lessons learned from that experience. Chapter 8 provides some basic conclusions, including advice from the study team concerning each of the three countries' needs for strategic government support of the energy efficiency agenda and about the roles of international financial institutions, as well as some operational suggestions for those countries and organizations considering new energy efficiency financing programs.

Part II of the book provides 13 case studies of different types of energy efficiency financing mechanisms that have been implemented in China, Hungary, Romania, India, Lithuania, the United States, Canada, Brazil, and Sri Lanka. The case studies describe advantages and disadvantages of the different approaches adopted, and specific lessons learned. They provide a platform for presentation of the synthesized conclusions in the main report.

THE NEED FOR ENERGY EFFICIENCY INVESTMENT FINANCING INTERVENTIONS

The critical importance of improving energy efficiency globally, but especially in rapidly growing developing countries such as China, India, and Brazil, is well documented in other analyses. IEA's *World Energy Outlook 2006* forecasts in its reference scenario a 53 percent increase in global energy demand with matching large increases in carbon dioxide emissions between 2004 and 2030 (IEA 2006b). China, India, and Brazil represent three of the top 10 energy consuming nations in the world now, and their share in total consumption will certainly increase. In the world as a whole, but especially in these rapidly growing developing countries, efficiency improvements to generate more economic output with less energy input is essential for reasons of energy supply security, economic competitiveness, improvement in livelihoods, and environmental sustainability. In an Alternative Policy Scenario, developed to investigate how more sustainable global energy supply and use might be developed by 2030, the IEA estimates that two-thirds of the hoped for carbon dioxide emission reductions in developing countries must come from improved energy efficiency, and the balance from changes in the mix of energy supply technologies.

To consider, specifically, how to achieve energy efficiency gains, the overall "energy efficiency terrain" must be dissected, as different aspects of the problem must be addressed in very different ways. At a basic level, reduction in energy use per unit of economic output can be achieved in two ways—through energy savings stemming from changes in economic structure, and through energy savings stemming from technical efficiency gains. Structural energy savings are the result of broad trends in economic development (for example, changes in sources of industrial value added) and are not very amenable to direct policy influence. Specific energy efficiency policies and programs, therefore, usually focus on achievement of technical savings—reducing energy use per unit of physical output, not output value.

When looking at technical energy savings potential, it is useful to separately consider new facilities and existing facilities. Improving technical energy efficiency in new facilities is especially important over the longer term, and especially in fast-growing economies. However, individual investors who build new power plants, transport systems, industrial capacity, or buildings must weigh many factors in deciding on technology and designs, and energy efficiency is only one factor—and often a minor one to them. The challenge for governments in this case is to influence the broad technology choice decisions of investors and encourage them to adopt energy efficiency solutions. The main tools that governments can use to intervene here are policy and regulatory tools.

When reviewing how to improve energy efficiency in existing facilities, it is important to further distinguish among different markets and types of projects to decide the most appropriate ways to intervene. Often, major energy efficiency gains can be achieved through investment in broad restructuring projects—to revamp entire production processes in industrial enterprises, or modernize urban transportation systems, for example. In these cases, too, energy efficiency is only one of many factors involved in the selection of technologies by investors, and the tools available to promote energy efficiency are again primarily policy and regulatory tools aimed at influencing those choices. In other cases, however, there are specific projects aimed at just improving energy efficiency—by replacing outdated boilers, utilizing wasted heat or industrial gases, or installing more efficient electrical equipment, for example. Here, development and financing of specific energy efficiency investment projects is required.

This book considers solutions for expanding investment in specific investment projects where the primary objective is to achieve energy savings. These investments represent only a piece of the overall required effort to improve energy efficiency, but it is the piece most amenable to specific energy efficiency investment interventions, as opposed to policy and regulatory actions. This book focuses on how to expand investment in the thousands of energy efficiency projects dispersed through economies, rather than those concentrated in a few very large companies, such as energy supply utilities. Energy saving opportunities can be found in existing industries and buildings of all types, in projects that typically range from US$50,000 to US$5 million in size. As documented in many other studies, a wealth of such "standard" energy efficiency investment projects remains unimplemented, especially in Brazil, China, and India, despite high financial rates of return and payback periods between one and five years (with many in the one- to two-year range). Capturing these project opportunities, which are often winners from the perspective of enterprises, investors, and society at large, has long been an attractive target. However, success has been elusive.

Success in capturing a bigger share of the large numbers of financially attractive energy efficiency retrofit projects has proven stubbornly difficult, primarily because the intrinsic nature of the projects and their broader setting make it hard for effective markets to develop naturally. In some countries, price distortions may undermine incentives, but in most sectors in Brazil, China, and India, and many other countries, this is not the case, as project financial returns are high in most instances. Flow of information about energy efficiency opportunities is far from perfect, but it has improved. In some countries, the required technical or managerial expertise is lacking, but in the case of Brazil, China, and India the issue is more how to bring existing strong expertise to bear.

Rather, the core of the problem in these and many countries lies in the intertwined problems of perceived high risk driving up implicit discount rates associated with projects, currently high transaction costs, and difficulties in structuring workable contracts for preparing, financing, and implementing energy efficiency investments. With their main financial benefits focused on savings of energy costs, these cost-saving projects rarely rank as equals with projects to expand production or capture new markets, especially in rapidly growing economies. Benefits in the form of calculated costs savings streams, as opposed to highly visible new production assets, appear as nebulous and inherently more risky to many. As project opportunities tend to be relatively small scale and dispersed, transaction costs can prove daunting unless mechanisms are put in place to take advantage of similarities among projects and bundle them. Some form of financial intermediation is usually required, unless enterprises use their own funds. Typically, therefore, implementation of energy efficiency projects involves interaction of both financing entities and technical experts with clients. Project delivery requires very efficient contracting to achieve this without driving up transactions costs—a challenge in any country, but especially where market institutions may be relatively weak, causing greater insecurities in contracting, as in Brazil, China, and India.

This poses two challenges for energy efficiency project developers. The first is to develop means to design, package, and finance energy efficiency investment projects efficiently, in ways which can overcome such problems in different local in-country settings. The second is to have those projects nudge forward changes in the local economic (institutional) environment so that energy efficiency investments arise spontaneously in the future—in other words, to "create" markets or to make markets for energy efficiency investment more "complete." Experience shows that these two processes seldom happen naturally at levels corresponding to more than a small fraction of the potential. Specific, customized efforts are required to develop investment delivery mechanisms that can operate sustainably in local markets and that can help expand local markets for various aspects of energy efficiency delivery services. This, then, is the primary focus of the agenda to expand uptake of financially viable energy efficiency investment projects.

DELIVERY OF ENERGY EFFICIENCY FINANCING IS AN INSTITUTIONAL DEVELOPMENT ISSUE

Development and operation of energy efficiency investment delivery mechanisms is an institutional development issue, and energy efficiency financing programs and projects must recognize this clearly. Lack of domestic sources of capital is rarely the true barrier; inadequate organizational and institutional systems for developing projects and accessing funds are actually the main problem. Therefore, mechanisms to capture the opportunities for energy efficiency investment need to be created or strengthened. This entails sustained efforts over years—new institutional constructs cannot be expected to develop and grow overnight.

Clearly, energy efficiency investment delivery systems must fit local institutional environments in order to be effective. This book finds that delivery systems developed in one institutional environment in one country often do not work effectively in a different institutional context. For success, local institutional environments must be well understood, and general solutions usually need to be at least partly customized for those environments.

Further, all energy efficiency financing mechanisms must successfully incorporate two functions: (i) a marketing, project development, and technical design function to efficiently package good projects; and (ii) a financing function. This book finds that a second common source of program failure is inadequate balance between these two functions, leading to insufficient project pipeline development to meet the needs of financiers, or inability to arrange and deliver financing for a series of well-developed projects. Both functions are discussed below, and in separate chapters of Part I.

Third, there must be sufficient incentives for the various players in a given energy efficiency investment delivery mechanism to undertake the functions expected. Again, while this seems like common sense, such incentives are at times difficult to achieve, given the variety of contractual arrangements that can be dictated by differences among local institutional environments. (The conceptual model presented in chapter 4 is targeted at these three fundamental requirements.)

Generally speaking, there are three basic types of investment delivery mechanisms for energy efficiency investment projects that have been popular in recent years:

- Loan financing schemes and partial loan guarantee schemes. These operate either within the commercial banking system or as specialized development agencies or revolving funds.

- Use of energy service companies (ESCOs). In this book, ESCOs are defined to include any company using energy performance contracting as part of energy efficiency investment transactions. An energy performance contract (EPC) in the ESCO business may be broadly defined as a contract between an ESCO and its client, involving an energy efficiency investment in the client's facilities, the performance of which is somehow guaranteed by the ESCO, with financial consequences for the ESCO.

- Utility demand-side management (DSM) programs. In DSM programs, energy distribution utilities organize all aspects of energy efficiency delivery, including financing, technical development, and interface with users.

It is common, also, to mix these mechanisms.

DELIVERING INVESTMENT PROJECT DESIGNS AND TECHNICAL APPRAISALS

For energy efficiency investments to be made, energy efficiency concepts must be marketed to enterprises, and specific projects must be identified, designed, and appraised. This requires marketing, project development, and technical assessment skill, typically provided by local energy efficiency experts. Human and organizational capacity is needed to define target markets and market outreach strategies, identify project opportunities, design appropriate project packages at end-user facilities, assess financial returns and the risks influencing delivery of the project cost savings cash flow, and understand the incentives to participate by each of the designated parties.

Early assessment of potential markets is important when developing energy efficiency delivery programs because different markets require different approaches. Selection of market segments for concentration will define organizational arrangements for technical work and the types of financial products developed. In addition, different stakeholders may have very different interests in market development strategies: one bank may be interested primarily in developing new small and medium-size enterprise (SME) clients, while another may be primarily interested in providing new services to existing large commercial customers.

Once target markets are defined, market outreach and marketing of project concepts needs to be conducted, followed by project development. Project development includes a series of key tasks including technical assessments, initial project identification and screening, customer enlistment and their acceptance of proposed project concepts, detailed design of project components, calculation of project economics, and identification and allocation of project risks.

Capacity to undertake project development work typically is found within project appraisal companies, energy survey and auditing firms, university or research institute departments, industry associations, equipment vendors, or ESCOs. In Brazil, China, and India, existing local capacity in the energy efficiency industry is fairly strong. In countries where local capacity is weak, development of this capacity then becomes a top priority—even a prerequisite—for energy efficiency project development. At times, capacity might be borrowed from neighboring countries, but experience has shown that excessive reliance on international consultants is generally unsustainable.

For countries such as Brazil, China, and India, the main issue is how to most efficiently access existing project development capacity. Almost always, both financiers and end users require some degree of independent assessment. For example, where a trusted ESCO might be able to fully meet the needs of both parties, usually the financier or the end user still wish to have some level of independent technical assessment. Choices then need to be made concerning the degree of outsourcing. Among end users, major industrial enterprises often may conduct technical assessments largely in-house, with perhaps only some very specialized expertise acquired from outside. Building owners, on the other hand, usually outsource nearly all of the project development and assessment effort. The situation among financiers also varies: some development finance institutions (DFIs) may have quite sophisticated in-house technical assessment capacity, whereas many commercial banks will contract out such work to trusted partners.

In all cases, energy efficiency investment financing mechanisms must include efficient and cost-effective organizational and institutional arrangements for delivering marketing and technical assessment requirements in which incentives of all the parties are properly aligned. In each respective economic environment, this is likely to include differing combinations of in-house expertise and outsourcing arrangements. Two points are worth special attention:

- The evolution of different available project development groups is often a critical factor determining their immediate effectiveness in a given energy efficiency financing scheme. Such groups typically have complex historical and staffing relationships that heavily impact their effectiveness as contractors for different financiers or end users, especially in developing countries.

- Keeping transaction costs reasonable is often a major challenge, especially given the relatively small size of energy efficiency loans. Design of programs to achieve this requires creativity and innovation. For example, for their general and energy efficiency lending to SMEs, Indian banks have relied on new geographical and industry-specific clustering approaches.

DELIVERING FINANCING

The main in-country options for financing energy efficiency investment projects include the internal resources of end users and outside sources of finance such as local banks (including local branches of international banks), leasing companies, and other nonbank financial institutions. ESCOs may provide financing to end users, but then they also will require financing from others. Other occasional sources include export credits, equity capital financing through special-purpose companies, financing from utilities repaid through energy bills, or informal sources. Multilateral development banks may provide direct financing to especially large end users such as utilities, but otherwise financing from these banks is usually channeled through local intermediaries.

Despite the variety of sources for financing energy efficiency projects, it is clear that ultimately the key source of sustainable and sizable flows of finance in most countries is the local banking sector. Although circumstances do vary considerably, the following observations hold true in many cases and are important in how banks tend to view the kinds of energy efficiency investment projects that are the primary focus of this book:

- Energy efficiency projects often represent a relatively small niche business for major banks.

- Project finance for projects targeting operating cost savings is nonconventional. Most lending in Brazil, India, and China is for working capital, and if project finance is available, it is usually only offered for large projects focusing on capacity expansion. Term lending for projects to improve business efficiency and increase productivity is less common.

- Banks lack knowledge of energy efficiency technology, and (reasonably) consider such specialized knowledge outside of the scope of their operational interest.

- Existing procedural frameworks within banks vary and banks are reluctant to alter them. To be operationalized effectively, new lines of business must be fit into existing systems.

- Customer relations are important, and the strategies of banks to attract and retain customers often dictate areas of interest in new business lines.

- Transaction costs for small and/or non-replicable projects are often a key issue.

In some countries, the local banking sector may be close to dysfunctional, the policy environment may be distorted, or the sector may be in the midst of major transitory reforms, making it difficult to use local banks for financial intermediation. Developing energy efficiency financing efforts may then involve difficult choices between incurring high risks of working in an immature banking sector, developing independent solutions, or foregoing the opportunity to achieve energy savings. If the program intervention is to proceed, especially with an independent approach, the high risks and needs for intensive efforts during implementation, including flexibility to adopt major midcourse corrections, should be recognized up front.

In many cases, energy efficiency projects can be attractively financed using existing bank loan products, without special adjustments or development of new financial products. However, modifications of financial products to match the characteristics of energy efficiency projects can help expand the market for such loans and increase uptake of financially viable yet unimplemented projects. The main direction for developing more customized financial products is to develop mechanisms that recognize and define the cost-reduction cash flow benefits of the projects and use this flow of funds as a source of loan repayment and security. The key is for financiers to increasingly recognize the characteristics of the cash stream generated by the projects financed and to structure loans and repayment assurances to best take advantage of that stream. There is an art to developing such enhancements and modifications of existing primary loan products.

Some of the special tools used by financiers to partially mitigate repayment risks from borrowers for energy efficiency projects, using the generated project cash flows, include the following:

- matching loan repayment schedules to project cost-saving cash flow
- use of escrow accounts for loan repayment, into which borrowers deposit cash from energy cost savings

- use of energy efficiency performance guarantees provided by third parties such as ESCOs
- use of ESCOs as project aggregators
- arranging for loan repayments to be made through utility bills
- development of build-own-operate or build-operate-transfer cogeneration projects under chauffage contracts.

MAKING INTEGRATED MECHANISMS WORK

For investment delivery mechanisms integrating project development and financing to be successful in increasing energy efficiency project investment, they should build upon the following principles:

- Delivery mechanisms need to be customized, based on a careful diagnostic review of the local institutional environment, including the financial sector, local capacities for technical assessment, the energy efficiency market, and the role of government. Such diagnostic review critically requires local expertise.
- End users should face commercial terms for the financing and technical services being provided as the best foundation for the creation of a sustainable energy efficiency market. End-user subsidies tend to ultimately undermine sustainable market development, because they are usually short-lived and can create market distortions and unrealistic expectations. However, concessional financing has often proven valuable to help buy down the high costs and risks of starting up new commercially oriented programs, build necessary new capacity, and assume risks with new approaches.
- Appropriate incentives must be included for the various actors in each mechanism to participate. Particularly important are incentives to generate deal flow. Combined with the previous bullet, this implies a focus upon organizational and institutional arrangements ("deal structuring") that deliver positive incentives

for all actors without relying upon long-term market-distorting subsidies.

Suggestions resulting from operational experience with the main types of energy efficiency investment delivery mechanisms are summarized below.

Energy efficiency lending through local commercial banks offers the highest prospect of program sustainability and large-scale impact. Experience suggests the following approaches:

- Design of major operations might best begin with partnerships with the financial intermediaries, and cater to their business approach and market development strategies. The financial intermediaries should select the operational arrangements for project development and technical assessment that best meet their needs and match their business preferences.

- Not all banks are likely to be interested in promoting energy efficiency projects as a specific line of business. However, an energy efficiency lending business may be useful for some as a means to achieve broader strategic goals. Some banks may be interested in developing such products geared to enhancing productivity as an extra service for existing good customers. Others may use energy efficiency loan products as a tool for entering or strengthening the bank's position in specific markets or business lines, such as the SME market or medium-term maturity lending to large industries.

- Integration of operational arrangements for technical assessment work with the financial intermediation of the banks is essential. Development and control of these arrangements would preferably be led by the banks.

Partial-risk loan guarantee programs supported by international financial institutions have shown some success in recent years in jump-starting energy efficiency financing programs through local financial institutions. This instrument is designed to defray part of the risks of loan repayment for energy efficiency loans. Such risks are often perceived initially to be high by local banks that are unfamiliar with energy efficiency business concepts or specialized means to mitigate those risks. The instrument also may provide a useful platform for delivery of a broad package of assistance to financial intermediaries. However, loan guarantee programs are not a broad panacea that can solve all the difficulties faced in efforts to expand energy efficiency investment. World Bank Group experience has shown that loan guarantees are especially useful where the banking system functions fairly well and the fundamental conditions that would allow energy efficiency lending to prosper are already in place.

Recent World Bank energy efficiency investment loan guarantee programs developed in Hungary and China show quite different approaches, although both have met with success so far.

The use of **DFIs and special revolving funds** is another common approach. An advantage is that DFIs and special loan funds can be designed as "one-stop shops," combining financial intermediation with strong project development functions, as the institutions have a dedicated, specialized purpose. In some cases where the local financial sector is under stress or in the midst of transitional reforms and restructuring, setting up special entities dedicated to energy efficiency lending may be the only way to establish funding channels. Their separation from the banking sector, however, also carries major disadvantages and major risks. DFIs are often established to act as catalyzing agents to pioneer the new business and help develop take-up by commercial banks, but this can create additional, difficult operational challenges. In some cases, especially with special revolving funds that have been added as components to bigger projects, capacities to deal appropriately with the details of proper credit evaluation and loan processing are often insufficient.

ESCOs can be an important market-based mechanism involved in the delivery of energy efficiency investment. ESCOs that provide financing to clients may be viewed as a partial financing mechanism for energy efficiency investment, operating at the retail level. These ESCOs serve as project aggregators, to which financial institutions may provide financing for a package of projects, thereby reducing their own direct involvement with end users. The mixed experience with ESCOs in developing countries suggests the following lessons:

- The ESCO model is not a magic bullet and does not solve basic problems of delivering energy efficiency project financing. Even when ESCOs provide financing to clients, the ability of the ESCOs themselves to obtain project finance is a central, difficult issue. The success achieved with ESCOs to date in China shows new ESCOs can play an important role if local institutional environments are suitable, but ESCO industry start-up is very complex, requiring complex contractual arrangements, staff with technical and financial and business experience, access to funding, and so forth.

- Long-term financing of ESCOs should be considered up front in any serious effort to promote local ESCO businesses. Programs that provide only technical assistance to build ESCO capacity alone have not proven very helpful in delivering large-scale sustainable impacts.

- Active government support for ESCO development is critical, especially in the early stages, as experience from both North America and China shows. This may include direct strategic support and/or assistance through market creation.

- The choice of ESCO business model should be determined by the local market, especially the choice between shared savings or guaranteed savings energy performance contracts. For some ESCO clients, such as building or commercial facility owners with little knowledge of energy saving technologies and their operation, the guarantee of energy savings may be very important. Clients in industrial facilities, on the other hand, may be very knowledgeable about energy savings of different investments and instead be interested in off-balance-sheet financing through ESCOs.

Utility DSM Programs. Although DSM programs were not one of the topics explicitly covered under the Three Country Energy Efficiency Project, these programs do represent another important option for promoting energy efficiency investments. DSM programs rely on the financial, organizational, and technical strength of major utilities to deliver numerous small-scale energy efficiency investments, using the relationships of utilities with consumers. In principle, the combination of delivery of energy efficiency together with delivery of energy supply would result in providing energy services as efficiently as possible. However, energy efficiency per se runs counter to the general business interests of supply utilities, since a kilowatt-hour saved is a lost sale and reduced revenue. Thus, government or industry regulators usually must provide special incentives to utilities to pursue such programs when the programs have the effect of cutting the utilities' revenues. Such regulation is difficult to undertake efficiently, especially in developing countries and emerging market economies. Under these circumstances, utility DSM programs may best be promoted only (i) where the utility industry is relatively responsive to public sector mandates; (ii) when energy efficiency efforts are combined with power-factor correction or load-management efforts that are in the financial interests of the utility; and/or (iii) in certain cases where promotion of energy efficiency may provide major benefits to the utility, such as expanding its customer base or reducing sales to customers whose tariff is lower than the cost of service.

MOVING AHEAD

One clear message from the experience of the Three Country Energy Efficiency Project is the importance of establishing and maintaining practical, operationally focused dialogue between the banking community and the energy efficiency practitioner community. This dialogue helped generate new energy efficiency lending programs in a number of Indian banks, laid a platform for the proposed development of a new energy efficiency financing initiative with major Chinese banks with World Bank support, and fostered the development of a new ESCO loan guarantee program in Brazil. Each country hopes to continue to build upon the platforms created after the close of this project.

Another clear conclusion is the central importance of strategic government support to more aggressively promote new energy efficiency financing mechanisms in each of these three countries. China's government has set an ambitious target to reduce energy use per unit of GDP by 20 percent during 2006 to 2010—and the challenge for the government is to mobilize effective implementation measures across the energy efficiency terrain. In Brazil and India, the study team recommends new, strategic reviews at the national level to consider medium- and long-term strategic priorities to improve energy efficiency. In the area of energy efficiency investment financing, a number of promising concepts have been developed in these countries, and it is important for the central governments to use their convening power and certain strategically focused but sustained institutional development support to enable new concepts to gain stronger operational footholds and to scale up initial experiences.

Well-targeted support from international financial institutions (IFIs) also can play an important role. The ability of IFIs to combine investment financing and project development support in multiyear packages is important in order to not just plan and train, but to implement promising new ideas. The IFIs also are able in principle to maintain a sustained presence, which is necessary to provide continuing support for new operational mechanisms from the design stage, through development and start-up, and finally operational rollout. However, because the problem to be solved is lack of adequate delivery systems for energy efficiency investment, and not lack of in-country capital, the success of IFIs should be measured in terms of energy efficiency results where possible, and not volumes of IFI lending, which is not directly relevant.

Project support from the Global Environment Facility (GEF) for commercially based energy efficiency financing programs has been especially critical and beneficial over the last decade. When introducing and developing new mechanisms, GEF grant financing for technical assistance and for investment support has been a critical tool—for trying new pilot projects, for covering part of the initially high transaction costs of schemes, and especially for helping defray initial risks. Continued strong support from the GEF can make a very big difference to the rate of success of developing and emerging market countries in this area in the coming years.

The authors hope that the analytical framework provided in this book and details concerning project implementation experience will be a useful contribution to countries considering development of specific new projects. Summarizing, the three biggest causes of operational failures in energy efficiency financing projects are (i) mismatches between the solutions attempted and local institutional environments; (ii) lack of proper balance between development of financial intermediation functions and project development functions; and (iii) lack of sustained effort and follow through, especially for adjusting institutional mechanisms and approaches during implementation in response to market changes or arising operational inefficiencies. To avoid these mistakes and to direct concerted efforts to achieve the best results possible in the future, the study team has the following broad suggestions:

- Careful diagnostic work on existing in-country conditions should form the basis for project design and interventions that fit within local institutional contexts.

- For projects involving financial intermediation, parallel attention is strongly recommended to (i) the details of developing capacities and mechanisms for financial intermediation, and (ii) project pipeline development and technical appraisal.

- It is important to incorporate periodic review and flexibility within the project design, so that programs can be adjusted during implementation.

- All of the above result in exceptionally high labor intensity for program management, operation, and technical support, not only during preparation but also during program implementation. High-quality and concentrated attention from program management and expert personnel is essential for new institutional mechanisms to be nurtured to success.

Energy efficiency financing operations are relatively costly and time-consuming to develop and implement. Development of the associated new institutional mechanisms requires intensive, multiyear efforts. If it is not possible to organize such efforts, it may be best to not attempt such ambitious programs. However, where possible, these programs can make a major, positive difference. With strong returns in terms of financial benefits to enterprises and energy consumers, and with very high potential returns per unit of public investment in environmental and energy security benefits to countries, further development of financing delivery mechanisms for sustainable energy efficiency undoubtedly has a major role to play in mitigating the energy development and climate change challenges of the future.

PART I

LESSONS FROM ENERGY EFFICIENCY FINANCING OPERATIONS IN CHINA, INDIA, AND BRAZIL

CHAPTER 1

INTRODUCTION

Energy—for heating, cooling, lighting, mechanical power, and various chemical processes—is a fundamental requirement for both daily life and economic development. But the needs of growing populations in their drive for greater prosperity are putting pressure on current energy supply sources and methods of energy use. Countries worry about the security of energy supply and rising costs. The negative impact on the environment of current energy systems is increasingly alarming, especially the global warming consequences of burning fossil fuels. The future requires change—through the development and adoption of new supply technologies, through a successful search for new, less resource–intensive paths of economic development, and through adoption of more efficient ways to use energy. Solutions are important for all countries, but not least for developing countries, where large populations have the greatest need for economic development and improved livelihood.

In the search for more secure and cleaner energy solutions, improving the efficiency of energy use tops almost everyone's list. Few will argue about the merits of achieving the same objective while using less energy, as long as it is cost effective. And opportunities for cost-effective energy efficiency improvements abound, especially in developing countries. But, capturing this potential has proved frustratingly elusive, again especially in developing countries and transition economies. Why is this so, and what can be done about it?

This book examines a series of programs adopted by various countries to increase energy efficiency investment, in an effort to shed light on what approaches seem to work best and why. As so many countries are now aiming to develop and expand programs to spur energy efficiency investment, the hope is that some insights can be gained from the past so that new efforts can achieve the best possible results. The book originates from a major program to exchange operational experiences in energy efficiency investment financing between the three largest developing countries—Brazil, China, and India—undertaken during 2002–2006 with support from the United Nations Foundation (UNF). This book's analysis is mainly drawn from the rich experience of these three countries, but also from a number of other country programs, especially those supported under World Bank Group projects.

ENERGY AND GROWTH

Global GDP is expected to more than double between 2004 and 2030. Accounting for a number of changes in the structure of global economic growth and technical improvements, the latest Reference Scenario of global primary energy demand completed by the International Energy Agency (IEA) projects 53 percent growth during 2004–30 (see table 1.1).[1] This growth implies a need for an additional 5.9 billion of tons of oil equivalent (btoe), which, if supplied through current means, will not only stretch the supply capacity of current systems but also entail global environmental consequences that are expected to be both very costly and damaging over the longer term.[2]

Eighty percent of the world's economic growth during 2004–30 is expected to occur in the non-OECD countries. As the developing countries seek to gain at least a modest level of prosperity, their energy demand is expected to almost double in the IEA Reference Scenario, adding 4.2 btoe to global demand. With the population of

Table 1.1. World Primary Energy Demand by Region, Mtoe (Reference Scenario)

	1990	2004	2015	2030
World (incl. bunkers)	8,732	11,204	14,071	17,095
OECD	4,518	5,502	6,261	6,860
Transition economies	1,488	1,077	1,259	1,420
Developing countries	2,612	4,460	6,372	8,619

Source: IEA 2006b.
OECD = Organisation for Economic Co-operation and Development.

Figure 1.1. Growth of Developing Countries' Energy Demand

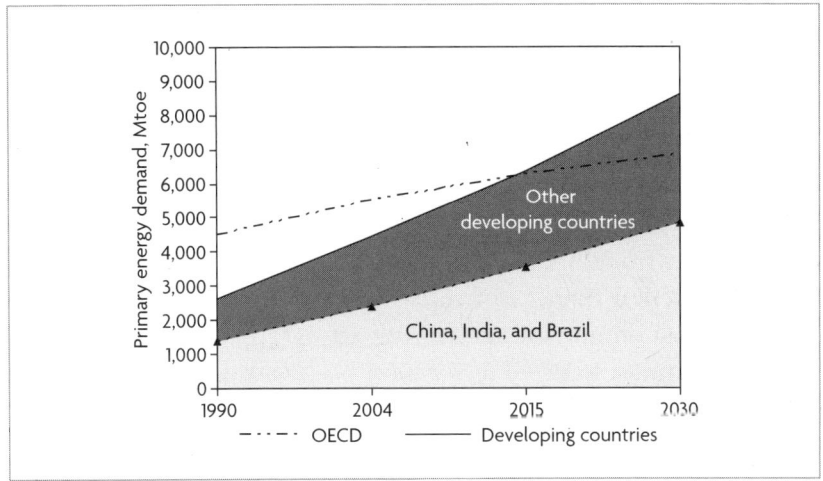

Source: Reference Scenario, IEA 2006b.

developing countries projected to be more than five times larger than that of the OECD countries in 2030, energy demand in the developing countries is expected to soon overtake total OECD demand (see figure 1.1).

China, India, and Brazil now represent three of the top 10 energy consuming nations of the world. As shown in figure 1.1, when combined, they account for well over half of developing country energy demand. China is easily the largest energy consumer of the three (4,769 mtoe in 2004), followed by India (1,103 mtoe) and Brazil (323 mtoe). Together, the three countries accounted for 40 percent of the world's population in 2004, a figure projected to fall slightly to 38.5 percent in 2030. In 2004, the three countries accounted for 22 percent of the world's energy use, and this is projected to rise in the IEA's Reference Scenario to 29 percent in 2030. This increase in energy demand alone accounts for 42 percent of the total projected increase in world energy demand over 2004–30.

In addition to the mobilization of resources to meet rising energy demand, a leading concern of most countries is ensuring their security of energy supply. Most countries are net energy importers, and rely on energy trade to obtain the mix of new energy sources needed, affecting their trade balances. A related concern is the prospect of sharp price increases and overall volatility in energy costs, stemming from tightening supply and potential disruptions in delivery.

Finally, concerns about environmental impacts of energy production and use are increasingly important. These include a wide range of local impacts, ranging from land and water use changes (for example, from development of hydropower, power lines, or coal or oil fields), to negative health impacts from increased air pollution (for example, from burning coal or operating automobiles). Regional environmental impacts of concern include increasing acid rain, and the most important global impact is global warming from increasing atmospheric concentration of carbon dioxide and other greenhouse gases.

It is now well recognized that global warming is occurring, and that it is largely caused by increases in the concentration of greenhouse gases in the atmosphere resulting from burning fossil fuels and other human activities.[3] The current excessive stock of atmospheric greenhouse gases was created primarily from OECD countries. New, harmful additions to the stock are now being created by all countries. With the rapid growth in energy use of developing countries, their additions of energy-related carbon dioxide to the cumulative greenhouse gas stock are projected to rapidly approach and then overtake the new greenhouse gas additions from energy use by OECD countries, according to the IEA's Reference Scenario (see figure 1.2). If the world is to effectively control the pace of global warming, all countries have important roles to play.

How to develop more sustainable energy supply and use systems is a hot topic. The impact of and role for developing countries are a key

Figure 1.2. Energy-Related CO$_2$ Emissions Growth to 2030

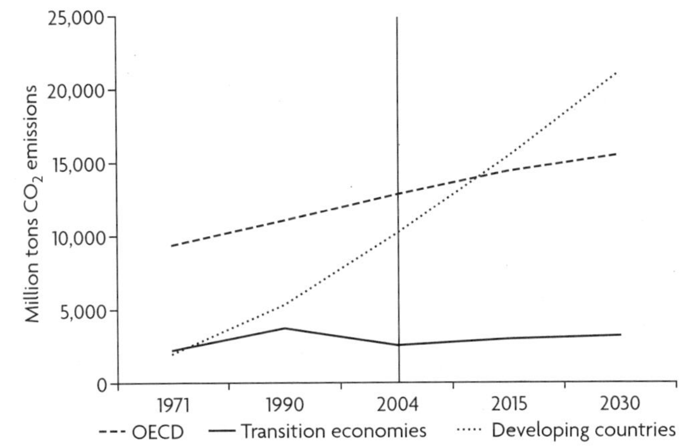

Source: Reference Scenario, IEA 2006b.
Note: Excludes emissions from international marine bunkers.

element of the debate. Can they forge new development paths over this century that can achieve better, more sustainable results than the wasteful paths set out by OECD countries during the last century without sacrificing their economic growth objectives? What is the cost of doing so, and, more importantly, what is the cost of inaction?

WHY ENERGY EFFICIENCY IS IMPORTANT

Improving the efficiency of energy use is a leading option to gain better energy security, improve industry profitability and competitiveness, and reduce the overall energy sector impacts on climate change. To provide a picture of how changes in the world's energy system might impact energy supply requirements and greenhouse gas emissions, the IEA prepares an Alternative Policy Scenario as part of its biannual World Energy Outlook (WEO). In the 2006 Alternative Policy Scenario, a series of more aggressive policies are modeled that reduce annual demand in 2030 by about 10 percent or 1.7 btoe. Increased adoption of energy efficiency measures account for a full two-thirds of the calculated reduction in global energy-related greenhouse gas emissions of 6.3 billion tons of carbon dioxide emissions when compared with the Reference Scenario. Other analyses use different assumptions concerning efficiency levels in "business-as-usual" scenarios and scenarios of more aggressive attention to reducing global warming, but yield similar conclusions on the important leading role of energy efficiency measures.[4]

As a domestic measure that reduces reliance on imported energy, energy efficiency programs are typically a key part of national efforts to improve the security of future energy supply. Energy efficiency is favored in environmental improvement strategies because it reduces the need for energy development, transportation and distribution, onsite use, and all the associated environmental impacts. But perhaps the greatest attraction of many energy efficiency measures is their cost effectiveness. Costs vary among technologies and countries where energy efficiency measures are implemented, but often are only one-quarter to one-half the comparable costs of acquiring additional energy supply (see the results from a recent survey of implemented energy efficiency investments in box 1.1).

Energy efficiency investments can reduce greenhouse gas emissions while also saving consumers money. For example, IEA estimated in its 2006 *Light's Labour's Lost* study that if end users were only to install efficient lamps, ballasts, and controls that would save money over the life cycle of the lighting service, then global lighting electricity demand would drop substantially and, as a result, total demand would be almost unchanged from 2005 levels by 2030 (IEA 2006a). This would avoid US$2.6 trillion in total expenditure on lighting through reduced energy and maintenance costs, and over the same time period avoid more than 16,000 million tons of carbon dioxide emissions.

Despite the win-win characteristics of energy efficiency investments both in terms of being cheaper than investments in energy supply and of reducing environmental impacts of energy use, it remains difficult to actually realize those benefits on a large scale. Many of the life-cycle cost savings measures are technically and logistically diverse, often small in scope, and typically do not compete well with opportunities for using up-front capital for capacity or market expansion. If left

> **BOX 1.1** Energy Efficiency Investments Are Very Cost-Effective
>
> A survey of 455 energy efficiency investments implemented in 11 industrialized and developing countries shows that the cost per unit of energy saved (present value over lifetime of the investment of 10 years) is on average US$76 per toe or US$11 per barrel of oil (in year 2006 U.S. dollars). This compares very favorably with the prevailing market price of energy, for example, more than US$60 per barrel of oil (in 2006 U.S. dollars). The figure below shows the wide range of cost effectiveness of various technologies. Still, more than 80 percent of the projects surveyed recovered their investment costs through energy cost savings within 30 months. Even one of the least cost-effective types of energy efficiency investments from the sample, in buildings, has life-cycle costs (8.6 U.S. cents per kWh over a 10-year lifetime) that are substantially below the costs that most final consumers have to pay for electricity. Not surprisingly, investments in countries such as India or China tend to be far more cost-effective than in industrialized countries.
>
>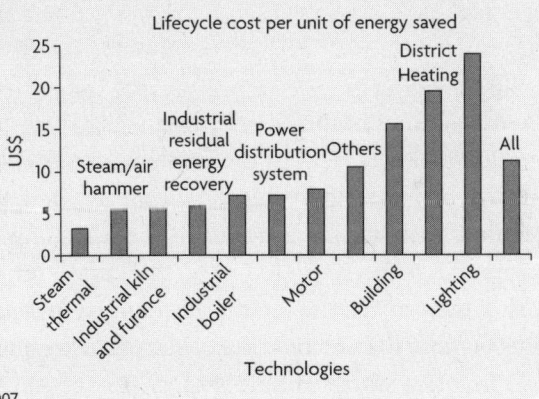
>
> *Source:* Shi 2007.

unaddressed, high transaction costs, perceptions of uncertain risks, and unmet needs for financial intermediation or outsourced technical input mean that much of the potential will remain unimplemented.

Multilateral and bilateral development agencies have supported energy efficiency activities in many developing and transition economies in an effort to move from pure technical assistance activities to demonstration and pilot projects and to scaled-up programs that move these investments into the mainstream. The World Bank Group—including the International Bank for Reconstruction and Development (IBRD), the International Development Agency (IDA), the International Finance Corporation (IFC), the Multilateral Investment Guarantee Agency (MIGA), and the World Bank-managed Carbon Finance funds—has been actively supporting energy efficiency, with over US$2.8 billion committed in direct investments, leveraged financing, and technical assistance for more than 100 projects in over 40 countries between 1990 and 2006.[5] The projects span all regions, but there is a significant concentration in Eastern Europe and Central and East Asia, and in a few economic sectors, in particular district heating and electric power. Despite this activity, there remains a huge amount of untapped potential, for reasons that are outlined in chapters 2 and 3 and discussed in increasing detail through the remainder of this book.

THE THREE COUNTRY ENERGY EFFICIENCY PROJECT

The Three Country Energy Efficiency Project was a major undertaking to share operational experiences in the implementation of energy efficiency investment projects between practitioners in Brazil, China, and India. This book is built largely as a synthesis of the many experiences shared by scores of these seasoned professionals working in these countries during the project, the lessons exchanged in discussion, and many thoughts on how to overcome the obstacles faced. Formally titled "Developing Financial Intermediation Mechanisms for Energy Efficiency Projects in Brazil, China and India" (also known as the Three Country Energy Efficiency Project or the 3CEE project), the project was funded by the UNF and the World Bank's ESMAP. It was implemented as a United Nations Environment Programme Project during 2002–06 by the World Bank and UNEP's Risoe Centre on Energy, Climate and Sustainable Development, in partnership with country teams of practitioners (also called core groups) in each of the three countries.

This project focused especially on how to build up energy efficiency investment project packaging capacity, both in existing financial institutions and through new entities, and to build the capacity of associated stakeholders. Activities included specialized technical assistance, training, and applied research covering the four primary areas of country interest: (a) development of commercial banking windows for energy efficiency; (b) support for developing energy service companies (ESCOs); (c) guarantee funds for energy efficiency investment financing; and (d) equity funding for ESCOs or energy efficiency projects. The project included a series of cross-country exchange activities that allowed energy efficiency practitioners from each of the three countries to learn from each other and to tackle jointly the practical problems in overcoming barriers to investments in increased efficiency. Country reports were prepared by each core group, providing in-depth discussion of the country framework, key related projects and relevant activities, and specific implementation experience and results of the Three Country Energy Efficiency Project.[6]

These project activities represented part of a larger World Bank program intended to set the stage for significant follow-up investment.

ROADMAP FOR THIS BOOK

The objective of this book is to provide a concise, operationally relevant review of lessons that have been learned from energy efficiency financing operations in a variety of countries in recent years. The following chapters address issues and lessons that go beyond those brought out in the individual country reports from the Three Country Energy Efficiency Project. The implementation of the Three Country Energy Efficiency Project itself provided the first exercise in action learning—an important element of energy efficiency project design that is discussed time and again in coming chapters. Among the more important lessons was that the differences between the three focus countries presented as important a learning experience as did their similarities. This led to the decision to increase the number and to widen the range of countries included in the case studies that make up Part II of this book. As the following chapters in Part I and the case studies in Part II demonstrate, the decision to increase the diversity of national institutional environments under study allowed the lessons from energy efficiency financing experiences around the world to be expressed not only with greater generality but also with greater confidence.

Chapter 3 offers an overview of why energy efficiency project investment opportunities in developing and emerging market countries are so widespread and the apparent returns so high. But in order to make clear which of those opportunities this book is directed at, first it is necessary to provide a new and different way of laying out the energy efficiency terrain which is done in chapter 2.

Chapter 4 integrates the logic of chapter 3 into a general model for designing and implementing energy efficiency financing projects and programs focused upon the types of energy efficiency investments discussed in chapter 2. Chapter 5 then deals with the marketing, project development, and technical assessment functions that must be fulfilled for any energy efficiency project, regardless of how it is financed. Chapter 6 discusses the different "mechanisms" for delivering financing for energy efficiency investments that have been used in recent years, and chapter 7 deals with the delivery process itself. Chapter 8 wraps up Part I by providing conclusions and recommendations that, like the remainder of the book, are directed primarily at government policy makers, management and staff of international organizations and financial institutions, energy efficiency practitioners, and key stakeholders from the financial community. The whole book is intended to assist this target audience in future efforts to develop workable energy efficiency financing mechanisms.

Part I and Part II of this book provide mutually reinforcing lessons from energy efficiency financing experiences in a number of countries. The case study environments range beyond China, India, and Brazil to include Canada and the United States, Hungary, Lithuania, Romania, Bulgaria, and Sri Lanka. The impact on delivery mechanisms posed by this rich diversity of institutional environments comes out both in the discussion in Part I and in the cross-country comparisons available in Part II.

Beyond national diversity, the case studies and the lessons drawn from them also present a variety of energy efficiency financing mechanisms and economic environments, ranging from loan guarantees to various forms of energy service companies and on-lending models. Still another dimension is the range of "deal structuring" arrangements that were developed in the process of making the various financing mechanisms work in each of the differing institutional environments presented by the countries involved in these studies.

NOTES

1. For this and other figures quoted from the IEA, see the IEA (2006b).
2. For a definitive analysis of the economic costs of global warming, see Stern (2006).
3. See International Panel on Climate Change (2007).
4. See, for example, Pacala and Socolow (2004), who analyze the potential contributions of a variety of measures for stabilizing atmospheric carbon dioxide over the next 50 years.
5. This includes GEF funding; see World Bank (2006a). For a review of the Energy Efficiency Portfolio; see World Bank (2004).
6. These country reports and numerous other project deliverables are available at http://www.3countryee.org

CHAPTER 2

SUMMARY OF THE ENERGY EFFICIENCY TERRAIN

Most global or country-specific reviews of energy efficiency focus on analysis of the potential for energy efficiency gains. They are important for establishing the priority of efforts to improve energy efficiency and the key areas with the most potential for improvement. These reviews typically begin with analyses of the energy intensity of economies or sub-economies, and then proceed with comparison of various unit energy consumption levels, analysis of the energy saving impact of different technical choices, and, in comprehensive studies, some analysis of the cost effectiveness of different choices and types of investment. Useful analyses of this type have been undertaken in many countries in the past, particularly in China, but also in India and Brazil, and can be readily found in the literature. In general, the greatest potential for cost-effective energy savings is concentrated in industrial energy use, residential and commercial building energy use, and energy use in the transport sector.[1]

A sense of the main areas with technical potential[2] for cost-effective energy savings then sets the stage for further questions. How can this potential be obtained? How and where can public policy be used to achieve the desired energy efficiency results?

Energy is only one of many inputs for production or basic living. As with any such input, appropriate pricing and use of competitive market forces are the usual broad economic policy measures to promote efficiency. In most cases, energy efficiency is only a small concern for those making investments—a host of other factors define competitiveness, or the attractiveness of equipment or buildings to purchasers. If energy prices broadly reflect costs, and competitive market forces largely prevail, should more be done? If so, what, and especially, how?

RECLASSIFYING THE ENERGY EFFICIENCY TERRAIN

To consider such implementation-focused questions, it is useful to classify the energy efficiency terrain somewhat differently from past analyses. The nature of energy efficiency issues and the measures to solve them differ not only by economic subsector, but also by the type of investment decisions involved. In particular, encouraging greater energy efficiency in the construction of new facilities (including industrial capacity, buildings, and transportation infrastructure) involves different issues and requires different approaches from those involved in promotion of energy efficiency renovations in existing facilities. Moreover, the appropriateness and nature of public efforts to encourage the most energy efficient choices vary substantially for different types of investment decisions.

Table 2.1 summarizes the principal interventions to improve energy efficiency and differentiates between new and existing facilities, as well as between key economic subsectors. The pages following then discuss the types of suitable interventions, first considering new versus existing facilities, and then considering each of the three main subsectors.

IMPROVING ENERGY EFFICIENCY IN NEW FACILITIES

In fast-growing economies such as India and China, the energy efficiency of new facilities is of paramount importance to planners when considering how to meet future national energy demand. At rapid

Table 2.1. Energy Efficiency Interventions by Economic Sector

Key Sector	Subsector	Principal EE Interventions
Industry	A. New plant	• Policy and planning • Equipment regulating standards
	B. Existing plant	
	1. Energy supply industries	• Restructuring investment • EE investment
	2. Industrial energy consumers	• Restructuring investment • EE investment
Buildings (commercial, public, and residential)	A. New buildings	• Building codes and standards • Policy and planning
	B. Existing buildings	• EE investment
Transportation	A. Motor vehicles	• Vehicle regulating standards
	B. Other	• Policy and planning • Restructuring investment

Source: Authors.
Note: EE = energy efficiency.

growth rates, forecasts clearly show how the energy efficiency of new industrial capacity, buildings, and transportation systems becomes the dominant factor in determining overall energy efficiency levels over the long term. Typically, the dynamics of fast growth also may compromise energy efficiency, as the pressure to build new capacity quickly and cheaply to overcome shortages or capture new markets often outweighs the consideration of long-term life-cycle costs, which lies at the core of investment in energy efficient facilities. Although of national importance in the aggregate, the efficiency of energy use is usually a minor issue at the micro, enterprise, decision-making level. Even for government agencies that may need to sanction investments, energy efficiency is at best only one of a series of concerns, including other environmental impacts, land use and zoning, local economic development and job creation, tax revenue implications, and infrastructure logistics. In China, this dilemma of new plant optimization is perhaps the greatest public policy issue today concerning energy efficiency.

To promote energy efficiency in new facilities, the tools available are policy and regulatory tools, including those listed in table 2.2.

Table 2.2. Typical Policy and Regulatory Tools to Promote Energy Efficiency in New Facilities

- Strengthening attention to long-term energy efficiency issues as part of the planning process, especially in urban and transportation planning
- Implementation of mandatory energy efficiency codes and standards for key types of energy-consuming equipment, motor vehicles, and new buildings (especially buildings where heating is required)
- Facilitation of energy efficient technology transfer, and adaptation and demonstration of key technological innovations in new plant in local markets
- Provision of information on the demonstrated performance and cost effectiveness of energy-efficient technologies
- Development of voluntary energy efficiency agreement schemes with industry, and energy efficiency benchmarking programs
- Provision of fiscal or direct incentives for implementation of new and innovative energy efficiency schemes
- Taxation of energy inputs (most governments are loathe to tax energy where trade competitiveness may be affected, but taxation of vehicle fuels is common)

Source: Authors.

Public financing may be required for policy and regulatory development and implementation, and potentially for certain concessional financing schemes to promote demonstration projects for new technology or other innovations. However, to promote energy efficiency in new facilities, the public policy focus remains on policy and regulatory tools, such as those described above. In promoting investment, the public policy objective is to influence broader investment projects where energy efficiency is only one, usually small, aspect. There is little public role for fostering mobilization of commercially oriented investment finance that is specifically and solely focused on energy efficiency for new facilities. The issue is not to encourage arrangement of commercial financing specifically for energy efficiency, but to encourage selection of energy efficient technology in the investments that are being financed.

IMPROVING ENERGY EFFICIENCY IN EXISTING FACILITIES

In existing industrial facilities, buildings, and transportations systems, sizable energy efficiency improvements may sometimes be gained purely through improved operational management, especially if previous incentives to reduce operating costs have been weak. Generally, however, the main gains in energy efficiency result from investment in technological upgrading and renovation. From the point of view of energy policy, the objective is to encourage cost-effective renovation investments which result in improved energy efficiency. However, appropriate public policy approaches will vary depending upon the nature of the renovation investments. When considering what public policy approaches to apply where, especially in industry, it is useful to divide renovation projects into at least two categories: (a) *restructuring, or major process/system renovation projects*; and (b) *standard energy efficiency projects* (see box 2.1).[3]

> **BOX 2.1** Why Distinguish Between "Restructuring Projects" and "Standard Energy Efficiency Projects"?
>
> Energy efficiency practitioners often argue as to whether or not the broad, multipurpose restructuring projects common in industrial renovation should be included in energy efficiency financing schemes, and if so, how. These projects can result in large energy efficiency gains, so why should they be distinguished from "standard energy efficiency projects" that have narrower objectives?
>
> A key reason to consider these projects differently is that their appraisal criteria are typically quite different. Industrial restructuring projects usually aim to improve an enterprise's overall competitive position in the market through major process change, change in product output, economies of scale, and so forth. Typically, restructuring investments are bigger and their payoff times longer than investments for standard energy efficiency projects. Project appraisal must consider the likelihood that the investment will result in a successful repositioning of the enterprise in the evolving market, and increased overall enterprise financial performance. Such outcomes require specific approaches and skills that have little to do with energy efficiency as such. In the 1980s, the IBRD financed a variety of such projects, some with a substantial energy efficiency focus. Results, however, were mixed. Public sector processes often proved out of step with the need for quick, ever-evolving adaptation to market change. With approaches and skills geared for private sector projects, the IFC has continued successfully with
>
> *(continued)*

> **BOX 2.1 (Continued)**
>
> industrial restructuring projects. However, energy efficiency gains are seen basically as co-benefits to broader improvements in enterprise financial performance, which is the primary focus of project appraisal—including how an enterprise fits into its particular competitive environment in terms of comparative advantage, product pricing, ability to adapt, and so forth.
>
> Appraisal of a standard energy efficiency project, in contrast, focuses primarily on ensuring delivery of the energy cost savings cash flow, which, by definition, comprises the main financial benefit of the project. Appraisal requires specific technical skills (see chapter 5). Financiers must be concerned about the financial health of borrowing enterprises, but the essential point is only to ensure that energy cost savings will indeed be used for loan repayment. Specific loan structuring techniques can be used to help ensure this (see chapter 6).
>
> The differences in objectives and appraisal criteria between the two types of projects means that the public policy tools to encourage energy efficiency through implementation of both types of projects are fundamentally different. For restructuring projects, the public policy goal is to encourage uptake of energy efficient solutions in these broader multipurpose investments. Therefore, what is needed are policy and regulatory tools that influence the broader decisions of others and provide incentives to maximize energy efficiency co-benefits. One practical and promising example being explored by IFC in several new projects is providing explicit incentives and technical support to banks involved in industrial enterprise lending to expand energy efficiency co-benefits through proactive promotion of energy efficient technology.
>
> For standard energy efficiency projects, public support is needed to help put effective project investment delivery systems in place to capture unrealized potential. Objectives and issues are clear, as described in most of the balance of this report. Including restructuring projects within specific energy efficiency financing schemes may or may not be appropriate, depending upon the case. What is essential is to ensure that the different due diligence and appraisal focus and skills required for these projects are brought to bear. Failure to properly review the broader issues surrounding restructuring projects is highly risky, and there are examples in recent experience of such dangerous failure in otherwise well-intentioned energy efficiency programs.
>
> *Source*: Authors.

SUMMARY OF THE ENERGY EFFICIENCY TERRAIN 41

Restructuring of Existing Operations
"Restructuring renovation projects" may be defined as renovation projects where the main objectives are not energy efficiency, but rather broader goals that often are fundamental to the enterprise's core business. However, these projects may produce the added benefit of substantial energy efficiency gains. In industry these projects may include major transformations in production processes, major changes in product quality or type, or expansions to capture greater economies of scale. Examples may include adoption of continuous casting in an older steel mill, re-optimization in a petroleum refinery, or adoption of new, high-speed production lines in a machine factory. Project benefits may include production of more profitable and higher-quality products, greater flexibility to meet market demands, and reduction in certain unit input costs, such as labor or raw materials, in addition to reduction in energy costs. In the transport sector, a project example is the revamping of urban public transport systems to provide better service and also reduce a variety of unit costs.[4]

Especially in the aggregate, and especially in industry, energy efficiency gains through such restructuring projects typically represent a substantial portion of potential energy savings at the national level. From an energy efficiency public policy perspective, it is useful to encourage restructuring renovation investment, where it is deemed cost effective. However, as discussed above for new facilities, these investments have broader objectives, and therefore the main energy efficiency focus involves encouraging the choice of energy efficient technology for investments being undertaken primarily for other reasons. Hence, the main public policy tools are the same policy and regulatory tools described in table 2.2.

Standard Energy Efficiency Projects
"Standard energy efficiency projects" may be defined to include renovation projects whose main objective is to increase energy efficiency and thereby reduce energy costs. One clear definition is renovation investment projects where energy efficiency benefits account for over one-half of the benefit stream. These projects tend to focus on renovation of specific equipment, building or factory energy service systems,

or energy-intensive processes. As described at the end of this chapter, standard energy efficiency projects have common characteristics, since they are investment projects for which the main benefits are a stream of future savings of energy operating costs.

CONCLUSIONS

Revisiting the paradigm outlined in table 2.1, three conclusions are summarized. First, the main tools for promoting energy efficiency in new facilities are policy and regulatory tools (including subsidy programs, if warranted) to strengthen incentives for investors to make energy efficient choices as part of broader investment decisions. Second, for major restructuring investments in existing facilities, the situation is similar: policy and regulatory tools are the energy efficiency promoter's best tools for influencing technology choice as part of broader investment decisions. Third, for projects focused squarely on the narrower goal of achieving energy cost savings, more direct involvement often may be warranted, which can facilitate the packaging and delivery of commercial financing for many financially viable projects that remain unimplemented.

The following sections briefly examine these conclusions for each of the three most relevant economic subsectors.

Energy Efficiency Investment in Industry

Broadly defined to include the energy and other utility service industries, industrial energy efficiency renovation projects in existing plant are a mainstay of the energy efficiency investment business in Brazil, China, India, and other countries with substantial industrial sectors. A wide variety of financially viable energy efficiency investment projects remain unimplemented in the energy industries, water service utilities, heavy metallurgical and chemical industries, building material industries, machine building and major light manufacturing industries, and small-scale rural industries.

When designing policy or program interventions to improve energy efficiency, it may be useful to consider energy efficiency investment in the energy industries (for example, electric power, coal, petroleum extraction and refining, district heating, and charcoal production) separately from other industries. Where energy efficiency implies reducing losses in production of the principal products of the firm, the core relevance of energy efficiency is easy to understand, and interest within firms is often high.[5] In such cases, the relevant technical capacities within enterprises also tend to be strong. From the perspective of multilateral development banks, an additional key differentiation is that financial intermediation is usually not required for energy efficiency lending to many energy industries, as these large, core infrastructure firms are common direct beneficiaries of multilateral bank lending. This makes energy efficiency lending much simpler for these banks.

Outside of the energy industries, attractive energy efficiency investment projects exist in both energy-intensive and other industries. A few typical examples include waste heat and gas recovery, boiler and steam system renovation, internal power system renovations, motor drive system renovations, cooling system upgrading, and a variety of specialized equipment replacements. Within the energy-intensive metallurgy, chemical, and building materials industries, some (but not most) energy efficiency projects may actually represent core production system restructuring—types of investments critical for the entire enterprise. Energy efficiency projects in industries that are not energy intensive, where energy costs may be well under 10 percent of production costs, are still an important part of potential energy efficiency gains at aggregate levels, and often have very attractive project financial returns.

Energy Efficiency Investments in Buildings

The distinction between new and existing buildings is essential in reviewing the building sector. The key components for ensuring energy efficiency in new buildings are (i) building design and orientation; (ii) ventilation and lighting system design; (iii) thermal integrity, including insulation and energy efficient windows and doors; (iv) proper construction methods; and (v) efficient heating, cooling, and lighting equipment. These components are the responsibility of designers, developers, and construction companies. Public policy should seek to encourage them to implement energy efficient choices. However, energy efficiency rarely figures highly in market-based building purchase decisions (compared with building cost, location, layout and appearance, convenience, and so forth). Utility bill cost reductions play out only over the long term, often for consumers who are different from initial building purchasers. Therefore, the main policy measure that many countries adopt—especially for energy efficiency measures that can be achieved with little additional cost—is mandatory building energy efficiency codes, at least for buildings with substantial heating requirements.[6] With a few exceptions, there is little role for public efforts to mobilize commercial financing for energy efficiency investments in new buildings.[7]

In existing buildings, the most lucrative energy efficiency projects involve renovation of energy service systems (such as lighting; heat, ventilation and air conditioning [HVAC]; and water pumping) in commercial and public buildings, including public health and education facilities and government offices. Similar system replacement projects exist in residential buildings, but are often more difficult to package attractively. However, the financial viability of major projects to improve the thermal integrity of buildings (for example, addition of insulation or replacement of windows) is highly site specific. Even in the best cases, investment payback periods for such projects are usually significantly longer than for efficiency measures in new buildings.

Energy Efficiency Investments in Transportation

Although the energy efficiency of transportation systems in the developing world is an increasingly critical issue at macro levels, there is limited scope for specific energy efficiency investment programs in this sector.[8] Modal choices have a major bearing on energy demand, but such choices involve broad transport policy and planning issues, including efforts to minimize many types of costs and maximize a variety of potential benefits. In the development and implementation of major transportation development projects, such as the development of new city bus systems or new urban traffic management schemes, it is useful to encourage choice of energy efficient technologies as part of project design. The same can be said for major transportation system renovation projects. In all of these cases, the challenge is how and when to introduce energy efficiency into the broader concerns of planning and project design. In some cases, where specific equipment is involved, such as motor vehicles, governments may usefully adopt energy efficiency standards to shape decisions of others. However, while there may be exceptions such as vehicle fuel conversions resulting in energy savings, there is not a major role for mobilizing financing for specific energy efficiency investments.

THE FOCUS OF THIS BOOK: STANDARD ENERGY EFFICIENCY PROJECTS

This book focuses on the problem of how to expand implementation of standard energy efficiency projects—that is, energy efficiency retrofit projects developed primarily to achieve energy cost savings in existing facilities. These projects are predominately found in the industrial and building sectors. As described in subsequent chapters, the challenge is to develop mechanisms that can efficiently package projects and deliver financing, as opposed to the development of policy and regulatory instruments to encourage energy efficient choices in broader investor and consumer choices.

Based in particular on knowledge of prevailing markets in Brazil, China, and India, standard energy efficiency retrofit projects do exhibit common traits, which are useful to consider when considering project promotion strategies:

- Investment costs usually range from US$50,000 to US$5 million per project. Larger projects tend to be restructuring projects that are not the subject of this book. The gains from smaller projects can have difficulty absorbing the transaction costs (further defined in chapter 3) of identifying, designing, and financing the project.

- Benefits occur predominantly as a temporal flow of savings on energy bills. Project energy savings actually achieved may or may not be easy to measure. One of the first functions of finance[9] is to provide products for transforming production and consumption across time. It is that function in finance that runs through all of the case studies and examples discussed in this book.

- The projects have high internal rates of return (IRR) in financial terms and quick paybacks (from less than a year to 4–5 years maximum, with notable exceptions being projects that improve the thermal integrity of buildings). Given that many projects with a high IRR remain unimplemented, there is little need to consider financing energy efficiency projects with a low IRR, especially in the face of the practical and the conceptual issues discussed in the next chapter.

- The projects tend to be highly replicable or capable of aggregation. In other words, similar projects can be aggregated and bundled into one financing package or can be replicated in a large number of similar enterprises or situations, with important implications for enterprise and/or lenders' transaction costs. Here also, there will be exceptions. Similar to small projects, unique projects pose problems in terms of the transaction costs versus the project-by-project gains to be realized. Aggregation and replication can be important for reducing transaction costs in many of the functions that must be accomplished when identifying, preparing, and financing energy efficiency projects.

The next chapter discusses why so many standard energy efficiency projects remain unimplemented, despite their financial attractiveness. Chapter 4 outlines an overall framework for how to address this problem through development of specific projects and programs.

NOTES

1. Although energy use for household cooking and hot water preparation is considered broadly under residential building use in this report, it may best be considered as an important separate category in many less-developed countries, especially if traditional fuels are included. The agricultural production sector, including direct energy use for irrigation and motive power, also should be considered in some cases, depending upon its relative importance and degrees of difference in the types of interventions needed.

2. In assessing energy savings potential at macro levels (for example, energy use per unit of value—GDP, industrial output, and so forth), it is critical to distinguish between energy efficiency gains from technical improvements and energy efficiency gains through changes in the structure of value added. Typically, changes in the structure of value added in the economy have a much greater impact on energy use per unit of value than technical efficiency improvements. However, the biggest structural changes may be at highly disaggregated levels (for example, more disaggregated than shifts between major economic sectors, or even between industrial subsectors), and therefore virtually impossible to measure. For example, if a men's shirt factory shifts production from simple worker's shirts, sold for US$2 each, to production of high-fashion shirts for export, sold for US$20 each, the energy consumption per shirt may change very little, but the energy per unit of value may fall to as much as one-tenth of that before. These types of structural energy efficiency shifts have been, and will continue to be, critical in shaping the energy intensity of industrial economies.
3. These concepts and definitions are developed in Ward, London, and Taylor (1994/2006).
4. In the building sector, the concept of "restructuring" projects for existing buildings might apply where existing buildings are gutted and fully renovated to achieve a variety of benefits. Typically, however, it is more cost effective to just replace existing buildings with new.
5. Examples include power loss reduction programs in power companies, district heating system renovations resulting in a reduction in heat generation needed for given heat service levels, system renovations in oil refineries to reduce own use and losses, and charcoal kiln upgrading. However, investments that improve the efficiency of nonproduct energy use tend to be like energy efficiency investments in other industries (for example, reduced electricity use in coal mine ventilation, or installation of more efficient electrical pumps in oil field operations).
6. In general, the principal-agent problem contributes to the difficulties in realizing energy efficiency investments in the building sector by limiting the effect of price signals. Public policy that improves the provision of information to end users, in addition to building and appliance codes, could lead to the uptake of energy efficiency investments in this sector. See Murtishaw and Sataye (2006).
7. One interesting exception is a business line among a few Chinese ESCOs. They are undertaking investments in service equipment that is incrementally more energy efficient (lighting or HVAC equipment) in new buildings. The ESCOs are receiving compensation from building owners based on energy cost saving performance. Another exception that could be supported

through public programs is the "energy efficient mortgage," a financing product that could help make new residential construction more energy efficient. See, for example, http://www.energystar.gov/index.cfm?c=bldrs_lenders_raters.energy_efficient_mortgage.

8. For a systematic overview of the policy options available to reduce the energy and GHG intensities in the transport sector, see Schipper, Marie-Lilliu, and Gorham (2000).

9. We return later in this report to other financial sector functions. This and the following analysis build upon Levine (1997).

CHAPTER 3

ORIGINS AND PERSISTENCE OF ENERGY INEFFICIENCY

Energy efficiency project market surveys in most developing countries and countries with economies in transition invariably find a series of energy efficiency investment projects with high financial rates of return (for example, over 20 percent), which remain unimplemented.[1] These projects are often classified as "win-win" projects: they are financially beneficial to the enterprises involved, and also beneficial to society's environmental interests. Why do so many of such projects remain unimplemented in countries where markets are generally working satisfactorily?

Led in particular by the programs and analytical work of the Global Environment Facility (GEF), work on this topic during the 1990s moved to systematic identification of various barriers to the efficient functioning of markets associated with energy efficiency investment. This was followed by design of initiatives to try to overcome these barriers. The GEF's very active Operational Program Number 5 (OP 5) has been built to try to catalyze efforts to remove the identified barriers.[2] After almost a decade of practice, "generic barriers" outlined in OP 5 provide a good summary of the main impediments commonly found to uptake of win-win energy efficiency retrofit projects. They include the following:

- lack of information
- lack of trained personnel or technical or managerial expertise
- below long-run marginal cost pricing and other price distortions (in some cases)
- regulatory biases or absence
- high transaction costs
- high initial capital cost or lack of access to credit
- high user discount rates
- mismatch of the incidence of investment costs and energy savings
- higher perceived risks of the more efficient technology

Energy pricing that at least roughly reflects costs, and access to information, technology, and technical and managerial expertise at reasonable cost, are basic requirements for markets associated with energy efficiency investments to function. With a few key exceptions (such as the rural sector in India), energy pricing is not a leading barrier in Brazil, China, and India, where project returns based on current retail prices are typically robust in most sectors. Flow of information about energy efficiency opportunities is far from perfect, but it is improving and is increasingly sophisticated. Technical and managerial expertise is relatively well developed, even if there remains room for improvement, especially at local levels.

In these three large countries, rather, the key impediments to effective energy efficiency investment through the market are the intertwined problems of current high transaction costs; perceived high risks driving up the implicit discount rates associated with projects; and difficulties in structuring workable contracts for preparing, financing, and implementing energy efficiency investments. This is also true in a host of other middle-income developing countries. Why are these problems so persistent and what can be done to overcome them?

THE IMPORTANCE OF INSTITUTIONAL ENVIRONMENTS FOR ENERGY EFFICIENCY INVESTMENT

In the final analysis, the key group of barriers (including high transaction costs, perceived high risks, and difficulties in energy efficiency deal structuring) are institutional in nature, meaning that they derive from the way that business interactions are handled and the potential conflicts and risks that such interactions can create.[3] With strong requirements for specialization, efficient packaging, and financial intermediation, the energy efficiency business is particularly dependent upon prevailing local economic institutions.

Over time, all societies tend to develop systems of informal and formal rules (institutions) that define acceptable behavior by individuals (including individual organizations such as businesses) and that also prescribe means of recourse when individuals are judged to not have followed the rules. The "formal" rules that are codified into laws affecting businesses ride on the shoulders of the "informal" rules that arise from long-developing customs and mores, yielding rich institutional environments (see Part II) that can be quite complex and that can differ greatly between countries and between points in time within the same country (such as countries in transition or "emerging" toward a market economy). Formal rules, laws, and regulations that do not comply with social customs can create great confusion and can lead to difficulties in enforcement of business agreements.[4] How both formal and informal rules are adjudicated or enforced is important: lack of reliability and effectiveness, or involvement of corrupt individuals or organizations, can create uncertainty among firms that see otherwise profitable opportunities for carrying on interactions and exchanges.

The effectiveness of the set of prevailing informal and formal rules governing operation of the market, and their adjudication, are very important for energy efficiency projects. This is because two core economic functions that are dependent upon the strength of prevailing market institutions are usually critical for efficient energy efficiency investment: (i) outsourcing governed by contracts to allow sufficient specialization, and (ii) deep and efficient financial markets for financing energy-efficient investments (including both initial and retrofit investments). Although these economic functions are important for many types of investment, the characteristics of energy efficiency investment projects, touched upon in the previous chapter, make them especially important:

- Unlike investments in fixed assets that yield productive output, such as renewable or other energy supply projects, energy efficiency investments aim to improve productivity through cost savings. Nonspecialist investors and financiers may be reluctant to part with upfront cash against the promises of future savings presented in calculations prepared by specialist project promoters.

- Projects are often small and scattered. Even if project rates of return are high, the overall financial significance of such small projects may be deemed unworthy of attention of busy individual managers, within either enterprises or financing institutions. If such small projects are not or cannot be aggregated to larger totals or managed (delegated or outsourced to specialized firms that can aggregate them across one or another dimension—see next bullet) in a way to economize on such transaction costs, they remain unfinanced and unimplemented.

- Technical content is diverse. Different technical solutions exist for hundreds of types of inefficient energy applications across the various economic sectors. Some technical solutions are fairly generic, and highly replicable among different consumers, but even in these cases, there is an art to efficient adaptation. Some technical solutions are complex and narrowly applicable—focusing on specific industrial process technology, for example. Although the industry-specific solutions often can be implemented by technicians within large or highly capable firms operating in highly competitive industries, the institutional environments of many developing countries and transitional economies discourage development of specializations in other dimensions than industry groupings—such as expertise in the art and the science of energy efficiency investment packaging and delivery—that can be made available across productive industries and sectors.

The strength of market-based economies is the potential they provide for economic gain arising from specialization and trade. In the market model, individuals and businesses specialize and develop "core competencies" within strategically defined tasks and then engage in market-based interactions and exchanges with each other to acquire things outside of their own core competencies. Economic models consistently show that the highest levels of economic welfare occur where the institutional environment allows exchanges to occur at low transaction cost (zero cost in pure theory models) and where all opportunities for mutually beneficial exchange are fully exploited.

When financiers use the term "transaction cost," they typically mean the administrative cost burden borne by each financing operation. This can include the (averaged-out) up-front costs incurred to get started a program of financing operations. Or it can include the loan-by-loan administrative costs for a line of financing operations. Obviously, high transaction costs of these types can eat away at the margin between the financier's cost of funds and gross return on funds (interest rate and service charges on loans, in the case of bankers). Thus, for example, high up-front costs to begin a small number of small lending operations would make no sense for a bank. Nor would carrying out large numbers of small operations that each yield so little gross margin as to fail to cover each respective loan's documentation and supervision costs. Of course, variations on these same transaction cost principles apply to energy efficiency end users as well.

Economists (starting from Coase; discussed further in chapter 4) have developed broader meanings of the term, encompassing also the costs associated with "information," such as searching for potentially profitable transactions, and for structuring deals so that the transaction can be controlled in the interests of the parties conducting the transaction (see the more complete definition in the Appendix).

This book will show that whether a deal can be done at all—and if so, how it is structured—depends very much upon local contracting and related economic institutions and the way these affect transaction costs, broadly defined. An important objective of Parts I and II of this book, then, is to demonstrate practical ways in which the economist's kind of transaction costs as they relate to energy efficiency can be reduced so that energy efficiency markets can emerge where they did not exist before or so energy efficiency markets can be broadened and strengthened where they previously had been thin and weak. The authors of this book believe this to be an important (new) role for public sectors and for development organizations to play—that is, undertaking projects and programs that help reduce transaction costs and make (energy efficiency) markets work better.[5]

In principle, the only difference between the financiers' and the economists' use of the term "transaction cost" is one of breadth. Both the narrowly defined and the broadly defined transaction costs have the same general effect: They eat away the margin of benefit over cost and serve as disincentives to transactions. At the same time, success on one front (for example, the narrowly defined version) can help progress on another front (transaction costs broadly defined), and vice versa. For example, the first bank's success in overcoming its own transaction cost problems and getting an energy efficiency finance program going provides encouragement and "models" that help reduce uncertainty (an aspect of the economist's transaction cost definition) for other lenders. This can play a valuable role in stimulating a broadening of the market for energy efficiency finance.

Of course, beyond the models and into the real world, the transaction costs of engaging in market-based exchanges vary greatly from one market to another and from one institutional environment to another. Where transaction costs of market exchanges are high, market exchanges tend to be few or nonexistent. That is, markets can be "incomplete" or "missing" in those areas, as with the market for risk management instruments or the market for energy performance contracts (EPCs; see definition in chapter 4) in many emerging market economies. Weak economic institutions tend to have a greater impact upon some types of exchanges than upon others. "Spot" exchanges, for example, can be more easily conducted where contracting institutions are weak than can exchanges encompassing more than the immediate point in time. Transactions taking place over longer periods (such as energy service contracts) require enforceable formal or informal agreements for assuring that the party receiving the deferred compensation is protected after a recipient has received goods and other benefits from up-front investment. Thus, the stronger the contracting institutions, the more numerous will be the longer-term contracts and the more "dense" will be the markets for those contracts.[6] This differentiation has a direct bearing both on the level of sophistication possible in the contracting for specialized technical services and on the development of the financial markets upon which energy efficiency financing relies.

Because financial markets are basically markets for contracts, the quality and workability of a country's contracting institutions will greatly influence the depth of that country's financial sector. On a more basic level, contracts are "promises," and the higher the probability of promises being kept, the greater will be the scope for creating and selling agreements (contracts) based on those promises. The foundational promise in financial markets (the first level of promises), of course, concerns government's reliability in maintaining the currency's value through time. Obviously, unreliable currencies create problems at the base of financial market development and increase both the need for and the difficulty in developing risk management instruments in the country's financial sector. Brazil, for example, has made substantial progress in controlling inflation in recent years, which is an important foundation for additional progress in financial market development. More work remains to be done in Brazil on the "third level" of promises outlined below.

The second level of promises involves government's reliability in repaying its own debts arising from deficit financing of public budgets—from which debt instruments in the more developed economies provide a "numeraire" (or comparison point) for gauging the risks implicit in other financial instruments.[7] The third level of promises, then, involves the country's contracting institutions discussed above and their impact upon the ability of enterprises and individuals to create and follow through on promises and agreements involving formal and informal contracts.[8] All three levels of promise-keeping play roles in the various indexes for comparing institutional environments and the scope for development of efficient and equitable markets in different countries, and are important factors in the development of financial markets and institutions of contracting that lie behind large-scale progress in energy efficiency.[9]

ADDITIONAL CHALLENGES POSED BY INSTITUTIONAL FRAMEWORKS IN DEVELOPING MARKET ECONOMIES

A range of sophisticated energy efficiency investment business models have been developed in various countries of North America and Europe where market and contracting institutions are relatively strong. Examples include chauffage arrangements, whereby customers buy energy services as opposed to energy resources; energy efficiency performance contracting (usually referred to more simply as energy performance contracting), and even wholesale trading of EPCs; and a variety of energy efficiency investment guarantee schemes. These models rely heavily on contracting institutions for outsourcing and on sophisticated financial intermediation. Even in these countries, however, where market economy institutions are quite strong, development of these business models has been a challenge. Public sector support was necessary to launch many business models and to push the market into new areas.[10] Effective business models also have typically taken many years to develop, scale up, and fine tune.

Countries with emerging market economies still under development pose additional challenges for the energy efficiency investment business. Table 3.1 shows that India and Brazil have some of the world's weakest institutions of contract enforcement, ranking them 173rd and 120th, respectively, out of the 175 countries evaluated in the World Bank's Doing Business database. By comparison, the United States and Canada, identified with success in developing contract-intensive forms of energy efficiency project implementation, are globally ranked 6th and 16th, respectively, in contract enforcement. China, at 63rd out of 175, has contract enforcement institutions at an intermediate level in the comparison presented in table 3.1.

Where the formal and informal institutions for ensuring promise-keeping are strong, not only can financial markets be deep and efficient, but opportunities for outsourcing tasks that can benefit

Table 3.1. Contract Enforcement: Brazil, China, and India Compared to Canada and the United States

Country	Procedures[a] (number)	Time[b] (days)	Cost[c] (% of debt)	Rank[d] (of 175)
Brazil	42	616	15.5	120
China	31	292	26.8	63
India	56	1,420	35.7	173
Canada	17	346	12.0	16
United States	17	300	7.7	6

Source: World Bank Doing Business Database (2006) at http://www.doingbusiness.org.
a. Number of procedures from the moment the plaintiff files a lawsuit in court until the moment of payment.
b. Time in calendar days to resolve the dispute.
c. Cost in court fees and attorney fees, where the use of attorneys is mandatory or common, expressed as a percentage of the debt value.
d. Country's place in the league table rankings amongst the 175 countries rated on this measure.

from specialization can be great. Thus, strength or weakness of a country's economic institutions can affect economic structure in two ways that are important to energy efficiency project development and financing. First, weak institutions of contracting (indeed, of property rights in general) will lead to financial sector limitations, as discussed above and in a number of studies cited below. Secondly, weak contracting institutions will lead firms to vertically integrate their operations, either carrying out tasks in house that in more developed economic environments would be carried out via market-based exchanges, or refraining from pursuing various tasks altogether, due to lack of in-house capacity.

Thus, limitations in the sophistication and reliability of contracting, as well as related financial sector institutional weaknesses, add to the difficulty of fostering robust energy efficiency investment in Brazil, India, and China, as in most developing and transition countries. Emphasizing the importance of institutional issues in energy efficiency finance in these countries, the following refinements can be added to the generic barriers identified in OP 5:

Missing or incomplete markets, in particular markets for risk. Immature or weak financial sectors in emerging market economies are not likely to have highly developed markets for futures, options, and derivative instruments—or to provide easy access by domestic investors to such markets abroad. In the absence of financial markets for risk, investors in "real sector" projects tend to load risk into the discount rates used in assessing all investments—new facilities as well as retrofits. This is a mathematical statement of investors seeking lower present costs (that is, initial investment) in exchange for higher future costs (for example, higher energy use) as a means of minimizing the risk of incurring sunk (stranded) investment costs.[11] Missing and incomplete markets for risk[12] will generate a large portfolio of potential retrofit projects in the real sector with higher-than-expected rates of return, particularly in those economies with the most poorly developed financial sectors. Unfortunately, the same mathematics of risk discounting will tend to affect the retrofit projects as well (unless major economic change in institutions and financial markets has occurred since the original new facility investments were made).

Political and economic uncertainty. Political and economic uncertainties in emerging country economic environments further increase the problem of loading risk into discount rates. Such uncertainty is often intertwined with weak contracting institutions (below) and with missing and incomplete markets for risk (above).

Weak contracting institutions (legal systems) result in insecure contracts with low certainty of equitable enforcement. Besides ruling out financial engineering involving both complex and simple risk management instruments, institutional weaknesses impede the use of "market" contracting (outsourcing) and relegate firms to the use of "hierarchy" (command-and-control of resources inside the organization), thereby impeding the application of specialized outside expertise in the solution of energy (and other) inefficiencies within an enterprise. Coasian analysis[13] converts the lack of knowledge of technical alternatives discussed in OP 5 into a problem of *where* the relevant knowledge is located and of the difficulties that weak contracting environments and underdeveloped markets present in bringing expertise (existent somewhere in the world) to bear upon the enterprise's problems. (Case studies presented in Part II illustrate alternatives in organizing ESCOs, for example, in the face of these and related issues.)

PRODUCTION VERSUS EFFICIENCY INVESTMENT IN RAPIDLY GROWING ECONOMIES

As discussed in this and following chapters, markets for "production" tend to be much better developed than the respective markets for "resource savings" (productive efficiency). Our studies suggest that this comparison constitutes a truism across a broad range of countries, from least-developed to most-developed (as the Canada-U.S. Case Study 8 in Part II illustrates). This does not mean that targeted interventions cannot be designed that will close this particular gap between the markets for production and for resource savings, but it does mean that the design of development interventions should pay particular attention to this general tendency. In fact, it is this tendency that makes growth without efficiency so potentially damaging to the environment and so stressful for (energy) resource supplies.

Growth Trumps Productive Efficiency

Particularly in the context of rapidly growing economies such as China and India, management attention tends to focus upon growth of sales, capture of market share, and achieving strategic targets such as first-mover advantage. Investment in higher production efficiency, which may pay off handsomely only over the long term, may often take a back seat. In addition, markets and opportunities tend to shift quickly, adding to risks in up-front investment that needs time to bear fruit. In essence the opportunity costs of capital and discount rates of medium- and long-term investment are often perceived, rightly or wrongly, as very high, and decision making may be almost obsessively focused on the short term.

The focus on production growth in these environments, when also coupled with weak contracting institutions, almost ensures that new facility investments will continue to include inefficiencies in energy use as well as in other dimensions of productive efficiency. Hence, the energy efficiency problem persists, with new facility additions unfortunately also adding to the stock of win-win energy efficiency retrofit projects that remain unimplemented. As discussed in chapter 2, governments seeking to improve the energy efficiency of new facilities need to look especially toward regulatory and policy tools.

The next and subsequent chapters discuss how effective investment delivery mechanisms can be developed to address the institutional barriers holding back investment in the standard energy efficiency retrofit projects in existing facilities. These delivery mechanisms are essentially deal-structuring innovations that are designed to address the specific institutional barriers at hand. The complexities and difficulties of structuring energy efficiency deals in these various and subtly different institutional environments, however, must be recognized up front. The energy efficiency investment delivery mechanisms need to be custom-built to meet the specific local institutional frameworks in place. As discussed above, since the current stock of unimplemented projects derives in large part from weaknesses in existing local market institutions, efforts to overcome these weaknesses must of course cater to the framework that caused the problem to begin with. In addition, although development of workable mechanisms to finance energy efficiency investments cannot simply stop and await the broader, long-term development of strong market institutions in the emerging market economies, this development should be done in such a way that it enhances (or at least does not impede) that broad evolution.

NOTES

1. One systematically analyzed example of such a survey in China can be found in Ward, London, and Taylor (1994/2006).
2. It should be noted that almost two-thirds of the operational programs and projects presented as case studies in Part II received support from the GEF under OP 5. OP 5 can be found online at http://www.gefweb.org/Operational_Policies/Operational_Programs/OP_5_English.pdf.
3. While the term "institution" commonly is used to refer to "organizations" (such as "the Bretton Woods institutions"), in institutional economics the

term refers to the rules and means for managing human interaction. See the glossary of terms in the Appendix.

4. As discussed in following chapters, this can pose particular problems when formal rules are "transplanted" from another country with its own complex institutional environment where the rules seemed to work fairly well. The often unfounded expectation is that these rules will work well in the new environment also.

5. Such an approach was proposed by Nobel laureate Kenneth Arrow as far back as 1969, in testimony before the Joint Economic Committee of the U.S. Congress, but has been slow to take hold in actual public sector and development organization operations.

6. "Depth" of financial markets, on the other hand, usually is interpreted as referring to the range of different instruments available in the market, which, of course, will also be affected by the quality of contracting institutions in a country. Research underlying these assertions is cited and discussed below and in following chapters.

7. Bernstein (1992) discusses, in a form readable by energy efficiency practitioners, the place of government debt in Modern Portfolio Theory, with U.S. Treasury instruments approximating the risk-free numeraire of Modern Portfolio Theory against which other risk-reward tradeoffs can be compared.

8. In an oft-cited article, Acemoglu and Johnson (2005) divide property rights (as defined in the World Bank's Doing Business database) into two components: (i) government refraining from expropriating private property, and (ii) government effectively and equitably enforcing contractual agreements between private entities.

9. Measures (often in the form of indexes) of the degree to which political and economic uncertainties translate into weak economic institutions and inefficient markets are provided by conservative think tanks such as the Cato Institute, Heritage Foundation, Fraser Foundation, and the Economic Freedom Network (http://www.freetheworld.com/release.html). The World Bank's Doing Business database, discussed further below, provides objective measures of business regulations and their enforcement (http://www.doingbusiness.org/).

10. For example, federal and local governments played a key role in the development of the energy performance contracting business, through initial market creation policies. See Part II, Case Study 8, on the development of the North American ESCO industry.

11. In other words, they apply a sometimes-crude form of "real options" in the choice of technology in energy-using as well as in other investments.

In simplest terms, real options techniques involve the creation of option points within real sector projects capable of imitating options available in the financial sector. See, for example, Amram and Kulatilaka (1999).

12. The concept of missing and incomplete markets derives from the Arrow-Debreu model of a competitive economy in general equilibrium with 'complete' markets, meaning a market for everything that affects human consumption-based welfare measures. In the presence of complete markets for risk, project discount rates would be determined totally by the pure (opportunity) cost of capital (that is, the risk-free rate of return in the alternative investment). Debates continue among economists over attainability of efficient general equilibria with incomplete *versus* complete markets, but that will not be our concern here. See Magill and Quinzi (1997).

13. This issue of market versus hierarchy was first identified by Ronald Coase (1937) and decades later made a central part of industrial organization research by Oliver Williamson (1985). See also chapter 4.

CHAPTER 4

MODELS FOR DELIVERING ENERGY EFFICIENCY INVESTMENTS

The conclusions in chapter 3 are founded on the presumption that making markets work better is an effective way to generate additional efficiency—whether it be in energy use, in pollution reduction, or in broader resource utilization. Provided that fundamental market operating conditions are in place (including pricing that provides suitable financial returns on economically viable energy efficiency investments), energy efficiency programs should use existing markets. However, where markets are incomplete or imperfect, efforts are required to design mechanisms to improve market functioning and analysts should work to build new institutional constructs that help to create markets where key elements are missing. These efforts are the focus of the GEF's OP 5 portfolio of projects in developing countries and emerging economies and of the various program and project case studies provided in Part II. All of these efforts aim to develop improved ways of delivering energy efficiency investment projects, in order to reduce transaction costs, perceived high risks, and financial intermediation system barriers.

While some success has been achieved, results in many of the new programs of the last decade have not been as great as hoped. At times, energy efficiency practitioners have been particularly frustrated with the seeming lack of progress. Officials ask why so many energy efficiency projects seem so cumbersome, slow, and difficult compared to other types of projects. What is the problem?

A big part of the problem is that energy efficiency projects usually must focus on introducing and developing new or modified ways of undertaking business transactions in order to overcome the mainstream barriers that stymie investment. For these new ways of structuring energy efficiency investment to work, they must be adapted to and accepted by the prevailing formal and informal institutional[1] environment. This is rarely a simple proposition.

In the effort to introduce energy performance contracting by ESCOs in China, for example (Case Study 7), one hurdle was to introduce to energy efficiency practitioners and energy users the basic new concept of deferred client payment of ESCO energy efficiency investment through performance contracts. But a bigger challenge has been to develop means for such business to operate efficiently and profitably within China's prevailing systems of legal, taxation, and contract enforcement regulations, conventions, and informal practices.[2] In the example of Case Study 5, Indian banks interested in energy efficiency programs for SMEs quickly saw that new types of outsourcing arrangements would be required to undertake specialized technical project identification and assessment work.

A key for ultimate success, however, will be how well existing contractual arrangements for outsourcing, and informal relationships behind them, can withstand inevitable challenges of uneven contractor work, costly delays, and so forth. Some Brazilian banks have expressed interest in participating in a new, publicly backed, guarantee scheme for energy efficiency loans. Among the thorny issues, however, is how to operate such schemes cost-effectively in the current structure and culture of the banks—for example, how to provide specialized assistance for technical and financial structuring quality control to the many decentralized bank branches that must be relied on for loan origination, while still keeping costs down. Countless other examples exist in Brazil, India, China, and elsewhere.

Experience in the three countries and elsewhere shows that, although difficult, specific and targeted institutional and organizational development efforts can successfully overcome barriers posed by current institutional environments, resulting in growth in energy efficiency investment. Part II includes case studies both of successful broad programs and successful small, narrowly focused efforts. The case studies also include examples that have been less successful so far in generating results.

A GENERAL MODEL FOR SUCCESSFUL DELIVERY PROGRAMS FOR ENERGY EFFICIENCY INVESTMENT

Integrating review of the operational experience and basic concepts of energy efficiency financing in the New Institutional Economics (NIE),[3] the study team draws several conclusions on the fundamental ingredients necessary for success in development of new energy efficiency investment delivery programs, and attempts to summarize these in a general model outlined below. Although they may seem obvious, it is surprising how often these general conclusions are overlooked in development project applications.

First, it is clear that three factors must come together for any energy efficiency retrofit investment to happen:

(i) Sufficient technical capacity must be brought to bear to identify, design, and implement the energy efficiency option (that is, knowledge of available options and the skills to draw upon them).

(ii) Financing must be in place to allow the intertemporal conversion of a flow of future savings into a capitalized investment in energy efficiency improvements.

(iii) There must be sufficient incentives for the energy-using enterprise, and every one of any other involved parties, to make the investment or contribute to the transaction as envisaged. Incentives are sufficient for a given party when that party believes that the probability of capturing gains surpasses all perceived costs, including transaction costs and opportunity costs.

One way to make energy efficiency investments happen is through a "hierarchy" approach: the energy-using enterprise undertakes the investment project by itself, with staff members inside the enterprise providing the technical capacity and the enterprise using internal funds for financing. This in-house solution is certainly common enough among large industrial enterprises in China, India, and Brazil. However, this solution alone continues to fall far short of meeting investment potential. Although vertically integrated organizations might do an acceptable job at handling routine, recurrent activities where in-house expertise can be kept working on a continuous basis, market-based exchanges (outsourcing arrangements) tend to be better at the specialized, one-off activity typical of energy efficiency retrofit investment. Hierarchy can perpetuate enterprise inefficiencies and/or substitute new ones, as enterprises draw on noncore competencies internally that are not as good as the same competencies in specialized firms. This is true not only for technical aspects, but also concerning finance, where use of internal financing foregoes the potential benefits of specialized structured finance that caters to the specific nature of energy efficiency projects. Moreover, although some large industrial firms may have capacity in-house to cover a certain share of the investment potential, maintenance of in-house capacity for energy efficiency becomes increasingly impossible further down the spectrum of firm size. Few smaller industries and building owners can justify such overhead. Finally, enterprise incentives to use their own funds may run thin, especially considering the high opportunity costs of use of such capital in rapidly growing economies (see chapter 3). Thus, huge parts of the market remain incomplete or missing.

To improve market penetration and investment efficiency, therefore, governments, international organizations, and energy efficiency advocates and experts seek to devise new institutional mechanisms that can provide the needed increased specialization. Each of the case studies of energy efficiency programs and projects in Part II describes an effort to develop new investment delivery mechanisms. This is the "world of boxes and arrows" that is familiar to energy efficiency specialists: a variety of schematically described solutions, depicting the various roles and implicit or explicit contractual relationships between different organizations that are collaborating to bring needed specialized resources/expertise and reduce transaction costs and perceived risks. In all cases, however, the three aforementioned requisites for success must be included in the designs: arrangements to bring sufficient technical capacity to bear, financing arrangements, and incentives for all players to play the roles assigned in the various schematic arrows and lines.

Most energy efficiency programs have been developed in an ad hoc way by project managers, business people, technical experts, and public officials. However, many of the basic principles concerning how industrial organization is impacted by the differences in transaction costs between conducting activities in house versus through contractual arrangement in the market were pointed out more than seven decades ago by Nobel economist Ronald Coase (1937). Later work by Coase (1960) spawned new and related research in transaction cost economics (see the discussion in chapter 3 and the definition in the Appendix) and in law and economics that by the 1980s had been brought to the core of industrial organization research by Oliver Williamson (1985). Recent work in the World Bank and in the economic development community is beginning to link these observations to institutional development issues in emerging market economies.[4] The concepts of institutions and of path dependence (see definition in the Appendix) of institutions suggest that the institutional environment of each country will evolve differently—creating differences between countries and within the same country over time, which have a major bearing on the specialization trends, contracting institutions, and their related transaction costs impacting industrial organization.

Based on both practical experience in energy efficiency programs, and backed up by broader analytical work, an additional key dimension then must be added to the three requisites already discussed for successful design of energy efficiency investment delivery mechanisms: the need to customize and adapt program solutions to match the multifaceted peculiarities of local business rules, customs, organizational behavior patterns, contracting and legal system realities, and organization of technical specialist capacities. Customization and adaptation of solutions is a theme that runs through all aspects of the energy efficiency practice. Accordingly, this important aspect is added to complete the generalized model for successful development of energy efficiency investment delivery mechanisms, presented in box 4.1.

As program designers in a given country begin to consider how to develop mechanisms to help capture a bigger part of the energy efficiency investment potential, they must build upon the existing systems and practices. How to best bring technical capacities to bear for a given program depends upon the specific technical skills required, what kinds of technical specialists are available where, how they are organized, how they may be contracted, and so forth (see chapter 5). Likewise, the sophistication, organization, and practices of banks and other financing entities also vary sharply among developing and emerging market countries. Equally important, but more difficult at times to grasp, are major variations in both formal and informal banking rules and procedures, banking culture, and the nature of the formal and informal financial contracting

---

BOX 4.1 Generalized Model for Developing New Energy Efficiency Investment Delivery Mechanisms in Developing Countries

Part I: Understand the **institutional environment** within which energy efficiency service transactions take place.

Part II: Pay careful attention to the **three requisites** that must be fulfilled within the respective institutional environment.

- Marketing/technical assessment
- Financing
- Incentives

Part III: Tailor the **institutional arrangements** for delivering the three requisites to the institutional environment within which the transaction is to take place.

Source: Authors.

relationships between financiers and customers. Whereas aggressive development of new financial products that can cleverly capture profits in energy efficiency investment might be of interest in some highly competitive markets, reliance on longstanding and informal "relationship banking" arrangements may work better in another setting (see chapter 6).

As the case studies demonstrate, the nature and direction of the incentives are very much affected by the institutional environment—the rules that govern individual and organizational action. Moreover, the effect of institutional environments upon incentives holds for all potential participants in energy efficiency transactions (and whether the energy efficiency transaction occurs through market or hierarchy), including incentives faced by the enterprise with the energy efficiency opportunity, by related financial institutions, by energy consulting firms, or by some form of energy efficiency project development company. For example, designers of the energy efficiency programs that focused upon public buildings in the United States and Canada (Case Study 8) had to understand the related public budgeting systems and the way the budget rules affect the incentives of agency managers who make the decisions about energy efficiency investments. In a similar vein, the Sri Lanka DSM program (Case Study 10) took advantage of existing load management incentives to enlist the electricity supplier in implementing household energy efficiency measures. Understanding—and sometimes changing—the rules affecting the project and market participants is part of the process that is discussed in subsequent chapters in this book.

By implication, then, the third element of the energy efficiency delivery model consists of the organizational (or "deal structuring") arrangements that must be tailored to the institutional environment within which the energy efficiency service transaction is to occur. These arrangements include such things as changing the budgeting rules facing agencies so that energy efficiency savings flow back to managers as incrementally spendable money. This seemingly simple incentive reiterates the extent to which these arrangements require specialized local knowledge.[5] Other cases illustrate the extent to which institutions, incentives, and risk are closely related and require joint consideration of the impact of institutional arrangements upon both risk and incentives. For example, the chauffage agreements undertaken by the Dongying Shengdong Energy Management Company (EMC) in China (Case Study 11) provides not only a good example of using existing markets to both reduce waste gas emissions and to convert those gases to usable products (co-generated electricity) but also of using deal structuring arrangements to reduce the risks felt by both sides of the energy efficiency transaction. The EMC reduces its own risk by retaining ownership of the equipment while reducing the client's risk by leaving the client free to draw power from the electricity grid in the event of a performance failure by the EMC. Practically every case study presented in Part II shows some form of organization/flowchart that describes the various parties and their responsibilities, with each set of responsibilities also implying a related set of incentives and risks that must be analyzed and dealt with as arrangements are being conceptualized.

Ideally, the project or program should not only come up with a way to structure a workable deal within the current institutional environment, but also it should work to improve the institutional environment so that (energy efficiency) deals will be spontaneously generated as simple market interactions in the future. Improving the institutional environment for energy efficiency investment can be addressed on a number of structured levels, all the way from reform of the judiciary to developing a new form of contract document that can be used as a model for a broader program (as with model contracts in the U.S., Canadian, and various other ESCO development programs). Included at this intermediate level is a loan guarantee program instituted in China (Case Study 1) that openly recognizes that project transactions are taking place in a transitional institutional environment which is likely to change in the future.

The tailoring of new mechanisms to fit local institutional environments should be recognized as a process, running through implementation, as opposed to just an aspect of program design. Institutional changes take time and require constant adjustments, which often can only be understood and undertaken as actual transactions are undertaken.

EXAMPLES OF DELIVERY MECHANISMS FOR ENERGY EFFICIENCY INVESTMENTS

Part II provides 13 case studies of energy efficiency finance delivery mechanisms that have been implemented in the recent past. The case studies are drawn especially from Brazil, India, and China, but case studies from Hungary, Romania, Lithuania, the United States, Canada, and Sri Lanka are also included. Most of the projects and programs in the case studies received some form of public sector support, and quite a few were the outcome of international donor projects. Generally, the case studies describe major programs involving development of mechanisms to deliver many projects, but there also are some cases of specific, innovative project financing structures that have been included to show interesting delivery mechanisms for just one or several projects. The case studies have been selected to show diversity in the types of delivery mechanisms that have been implemented, and some of the advantages and disadvantages of each. Common points among all of the cases include the following:

- All mechanisms focus on standard energy efficiency renovation investments in existing facilities.

- Only delivery mechanisms that have already resulted in investments are included (plans for new mechanisms that have not been implemented have been excluded).

- Out of necessity, each delivery mechanism has been adapted to the local institutional environment.

- All mechanisms incorporate considerations for provision of investment financing and for the technical aspects of project identification, development, appraisal, and implementation.

Generally speaking, there are three basic types of delivery mechanisms for energy efficiency investment projects that have been popular in recent years: (i) loan financing schemes, and partial loan guarantee schemes; (ii) ESCOs; and (iii) utility demand-side management programs. It is common, also, to mix more than one of these in development of specific investment delivery programs. In some cases, programs also include schemes for equity investment into special financing entities or ESCOs. Each of the three basic types of mechanisms is introduced below, and then referred to and analyzed in greater detail in subsequent chapters.

Energy Efficiency Loan Financing and Loan Guarantee Schemes
Six of the case studies involve development of specific energy efficiency loan programs. To be successful, however, each such mechanism must not only include provisions for loan finance, but also specific provisions for project generation and technical review. All of the schemes reviewed have the ultimate objective of fostering increased energy efficiency lending in the commercial banking system. However, public sector organizations play a key role in most of the mechanisms, to catalyze and help spark the business. Although there are other similar examples also under implementation, the following case studies are included in this review:

(i) an energy efficiency loan program undertaken by a publicly owned financial institution (IREDA, India Renewable Energy Development Agency)

(ii) an energy efficiency loan fund developed by a publicly owned, independent organization (the Romanian Energy Efficiency Fund)

(iii) an energy efficiency line of credit for thermal rehabilitation of multiresidential buildings (the Lithuania Energy Efficiency and Housing Pilot Project)

(iv) a series of SME energy efficiency loan programs developed by Indian Banks (Energy Efficiency Cluster Lending for SMEs by Indian banks)

(v) an energy efficiency loan partial risk guarantee program, administered by a multilateral bank (IFC's Hungary Energy Efficiency Loan Guarantee Program)

(vi) a partial risk guarantee program for loans to ESCOs, administered by a publicly owned guarantee company (the ESCO Loan Guarantee Program of the China National Investment and Guaranty Company, I&G).

In addition, a new Chinese program to develop lending businesses in a series of banks specifically for large-scale energy efficiency projects, to be financed in part by a US$200 million World Bank loan, which was a direct outgrowth of the Three Country Energy Efficiency Project, is discussed in the text but is not included in the case studies as it has yet to be implemented.

The implementation record of energy efficiency loan and loan guarantee funds to date is mixed. While a number of mechanisms designed for this purpose have operated effectively, achievement in most cases of the ultimate goal of rapidly expanding commercial bank lending for standard energy efficiency projects remains elusive. As described in subsequent chapters, key challenges are to develop mechanisms that (i) incorporate project development and technical review requirements efficiently; (ii) incorporate loan financing structures that can capture as much of the potential energy efficiency investment market as possible; and (iii) are aligned with the strategic, commercial objectives of the financing institutions, providing suitable benefits to them.

Delivery of Energy Efficiency Investment through ESCOs or Other Independent Actors

In addition to schemes that focus on development of energy efficiency investment delivery mechanisms implemented directly by financing institutions, another popular option is to foster development of independent energy efficiency investment "packaging agents" that can operate in the market. These entities become a focal point of project development and implementation—working with end users to identify and design projects, arranging financing, and playing a heavy role in project implementation.

ESCOs represent a relatively elegant energy efficiency investment delivery mechanism[6] that has long caught the imagination of energy efficiency practitioners. These companies play the packaging agent role, using energy performance contracts (EPCs) with end users. For the purposes of this book, an EPC may be defined very broadly as a contract between an ESCO and an energy user client whereby the ESCO provides specified energy efficiency investment project delivery services in exchange for remuneration that is in some way dependent upon the energy efficiency performance of the project delivered. These contracts can provide a form of guarantee both to end users and to financiers that the energy cost savings of the project will materialize as forecast.

The ESCO mechanism is a concrete, market-based means to address the delivery mechanism issues of energy efficiency investment, and as such is appealing on many fronts. The proposition is for groups of energy efficiency project specialists to identify and prepare unimplemented, high rate-of-return energy efficiency projects, and then package them for financiers, guaranteeing the performance of their work, and generating substantial profits back to the ESCO from the high returns of the energy efficiency projects. However, results from the many ESCO promotion efforts over many years are mixed at best. Arbitration between the various actors is complex, involving contractual challenges. Credibility and relationships need to be honed with end-user decision makers, financiers, technical staff, and technology suppliers. Last, but not least, a unique mix of skills is required, including skills in energy efficiency technology, project implementation management, financial risk management, and business development. When ESCOs are involved in providing some of the financing to end users, legal, accounting, and tax issues may provide additional hurdles, especially if energy performance contracting is not well understood in the market.

Three specific ESCO case studies are included in Part II, including (i) "full service" ESCOs in China, using shared savings energy performance contracting, and providing project financing as well as design, equipment procurement, project implementation, and user training services; (ii) the U.S. and Canadian ESCO business models for energy efficiency investments in public sector facilities; and (iii) ESCO project development in Brazil using electrical utility financing from Brazil's energy efficiency wire-charge.

Additional case studies in project finance, using third parties aside from financiers and end users as transaction packagers include (i) a standard build-own-operate chauffage arrangement for industrial cogeneration (example from China, using industrial waste-gas as a fuel); (ii) IQARA's operations in Brazil expanding gas usage through build-own-operate arrangements; and (iii) an equipment leasing arrangement with end-user payments made through electricity bills (example from an Indian capacitor leasing project).

Utility DSM Programs
A third major type of energy efficiency investment delivery mechanism is to utilize energy distribution utilities to organize all aspects of energy efficiency program delivery, including financing, technical development, and interface with end users, through "demand-side management" (DSM) programs. These programs were especially popular for electricity conservation in North America in the 1980s and the early 1990s as policy makers sought to encourage utilities to meet future system demand on a least cost basis considering both demand and supply-side options. The programs are mandated by governments, and rely on the financial, organizational, and technical strength of electric power utilities to deliver numerous small-scale energy efficiency investments, using their relationships with consumers. An appealing conceptual aspect is to put delivery of energy efficiency together with delivery of energy supply, in an effort to provide energy service on both sides of the meter as efficiently as possible. The major drawback is that energy efficiency per se runs counter to the general business interests of most supply utilities—since a kilowatt-hour saved is a loss of sale and sales revenue. Thus, government or power industry regulators must usually provide special incentives to utilities to pursue such programs when the effect of the program is to cut the power company's revenues. Such regulation is difficult to undertake efficiently, especially in developing and emerging market countries. As further discussed in chapter 7, utility DSM programs may best be promoted where (i) the power industry is relatively responsive to public sector mandates; (ii) energy efficiency efforts are combined with load management efforts that are in the financial and operational interests of the utility; and/or (iii) certain cases where promotion of energy efficiency may provide major benefits, such as expanding a utility's customer base (for example, for gas distribution companies seeking to expand connections) or reducing sales to customers whose tariff is lower than the

utility's cost of service (for example, public lighting or subsidized agriculture consumers).

Part II includes an electricity utility DSM case study from Sri Lanka (Case Study 10). The energy efficiency wire-charge in Brazil, mentioned above under ESCO case studies, has elements of a DSM program as well.

The Benefits of Small, Targeted Program Interventions

The discussion in Part I and the case studies presented in Part II should not leave the incorrect impression that only "big" energy efficiency projects and programs with "big" impacts on institutional change are useful. Nothing could be further from the truth. In fact, starting small, studying, and carefully documenting what is learned at each step before embarking on huge investment programs is much more consistent with this book's arguments about the way that sustainable organizational and institutional change actually seems to work in the energy efficiency context. The analysis in Part I leads to the conclusion that some of the programs discussed in Part II might have benefited from even smaller beginnings with even greater monitoring and evaluation as the programs went forward and got scaled up.

The Role of Monitoring and Evaluation

Introspection suggests another valuable lesson that comes from starting in a targeted area, testing, and learning before scaling up. It is critical to build into energy efficiency programs a formal process for learning and for integrating those lessons before moving to the next step. That was a core objective in the organization of the Three Country Energy Efficiency Project, with the regularly scheduled meetings and conferences playing a central role in spreading lessons amongst the teams working in the three countries. Of course, one of the valuable lessons coming from this approach was that the ways of implementing the various financing mechanisms differ much more among the three countries than had been anticipated when the Three Country Energy Efficiency Project was originally designed.

Making Energy Efficiency Investment Happen Spontaneously

A major lesson from the past is that externally identified, development assistance–funded energy efficiency investments will never "fix" all the existing energy inefficiencies by themselves. Much more can be done when such projects and programs are focused upon helping to develop markets that will spontaneously move toward finding energy inefficiencies and fixing them as profit-making activities, without a government official or an external expert ever having had to be involved or even to think about it. Often the smallest and most inexpensive program interventions can help that to happen. Again, model contract documents catering to local applications immediately come to mind, as does helping one major bank figure out how to make money off of energy efficiency loans. Private sector entrepreneurs are notoriously competitive and ambitious, universally, and if they see somebody else making money from something, they tend to immediately rush into the market—as indicated by the competition faced by IREDA (Case Study 4) as that organization worked to find ways to make energy efficiency projects profitable and workable. And much the same can be said about the spread of ESCOs in China, as the early efforts began to "make a market" for that form of energy efficiency business in that environment. So "big" is neither necessary nor sufficient for energy efficiency project or program success. This book suggests that starting with interventions that are targeted at areas showing the most potential in that particular time frame and institutional environment, studying and analyzing successes and failures from those early efforts, and recrafting along the way, often has been found to work much better.

NOTES

1. As mentioned in chapter 3, in this analysis, the word "institutions" refers to the rules and means for managing human interaction, as used in the New Institutional Economics (NIE). The link between institutions, finance, and economic development is a growing area of study among economists (see, for example, La Porta et al. 1998 and 1999; Rajan and Zingales 1998; Acemoglu and Johnson 2005).

2. Generally, a lack of transparency in the courts and in government agencies can induce firms (SMEs in particular) to rely upon relationship-based contracting, to eschew impersonal and arms-length agreements, and to use third parties of common interest to arbitrate contract disputes rather than the court system. Firms in India and China are heavy users of informal systems in both regards (see Allen, Qian, and Qian 2005; and Allen et al. 2006).

3. Terms used here and elsewhere in this report originating from the NIE are defined in the glossary in the Appendix.

4. Coasian theory (Coase 1937) suggests that low transaction costs *via* markets would promote de-integrated market networks in contrast to large, vertically integrated firms. Indeed, studies argue that weak contracting institutions and weak financial sectors tend to favor large firms over small firms (Banerjee and Newman 1993; Galor and Zeira 1993; Aghion and Bolton 1997; Greenwood and Jovanovic 1990), and that weak contracting institutions and underdeveloped finance reduce the growth rates of small firms (Beck, Demirguc-Kunt, Laeven and Levine 2005).

5. Sometimes, unfortunately, local knowledge can be further fragmented within the country. As pointed out by Fernandez and Kraay (2005): "... firms operating in city-industry cells with worse contracting institutions by one-standard deviation (i.e., where the functioning of the judiciary in business disputes is worse) are 14.6 percent more likely to belong to a business association [in other words, to use "informal" means of developing and enforcing agreements] than firms in a city-industry cell with (sample) average contracting institutions."

6. The pure, full-service ESCO model relies upon sophisticated contracting institutions and a highly-developed legal system. Variations of that model have provided mechanisms for generating, testing, and then mainstreaming model contracts and other means for encouraging improvements in contracting institutions and financing mechanisms in emerging markets. Related variations of the ESCO model in different institutional environments are discussed further in following chapters.

CHAPTER 5

IDENTIFYING AND DEVELOPING ENERGY EFFICIENCY INVESTMENT PROJECTS

For energy efficiency investments to increase, good investment projects must be identified and developed into project packages that are attractive to both project owners and financiers. Who does this work? What kinds of projects should be sought? What does project development entail? How can it be organized most efficiently, with minimum cost to all parties? The importance of addressing these questions cannot be overemphasized. One of the main reasons that some previous lines of credit developed for energy efficiency investment have failed is that project pipeline development systems did not work effectively.

Drawing in part on case studies in Part II, this chapter seeks to shed some light on these questions. A first issue that public agencies, commercial financiers, or both must consider when developing a new or improved energy efficiency investment delivery program is market selection and outreach. What types of projects should be the main focus, and how can reliable prospective clients best be enlisted? A second set of issues concerns the project analysis required for an investment to be bankable. Much, but definitely not all, of this analysis involves technical evaluation. A third set of issues concerns who might best undertake the necessary detailed project analysis. In some cases, local expertise is lacking. Often, however, the necessary expertise is available, but the key issue is how best to access it. Organization of local technical expertise varies strikingly between countries. Therefore, decisions need to be made by both end users and financiers as to what aspects of project development and assessment should be undertaken in house, and what aspects should be outsourced to what groups, in order to achieve the best results with minimum cost.

MARKET SELECTION AND OUTREACH

Market Selection

It is important to decide early in program development which markets to target initially. An objective assessment of demand and potential for a specific energy efficiency financing business will be essential for funds to be secured for active program development and for the backing of financiers. But early market assessment and strategic targeting is also important for reasons of operational strategy. Penetration into different energy efficiency markets requires different kinds of technical expertise, and different types of institutional connections, and hence, different program approaches.

Looking through a few examples from the case studies in Part II, substantial variations in target markets are evident:

- The Chinese ESCO industry (Case Study 7) has focused primarily on investment in predefined technical solutions, replicated among a variety of different clients. Substantial differences in business approach have evolved, however, between ESCOs focusing on industry and ESCOs focusing on commercial buildings, and only the most sophisticated ESCOs can accommodate both.

- The SME energy efficiency lending programs of the Indian banks (Case Study 5) define specific types of technical solutions for specific industries in specific geographic locations, and then seek to replicate a large number of basically identical small projects.

- The Romanian Energy Efficiency Fund (Case Study 3) has relied primarily on investment project proposals typically developed as customized projects to meet the needs of specific industrial enterprises and municipal utilities, through individual technical studies and enterprise audits.

- For its industrial energy efficiency programs, IREDA (Case Study 4) has invested in specific, customized technical solutions for individual industry clients, but with a view to then expanding their loan portfolio to similar clients who propose similar investments.

- The Lithuanian Energy Efficiency and Housing Pilot Project (Case Study 6) has focused primarily on solutions for urban residential housing blocks, to develop a limited number of narrow energy efficiency solutions as part of a broader integrated program to improve urban heating efficiency.

Most energy efficiency programs differentiate their market focus to some extent based on project size and technical sophistication. Focus on small projects, many of which may involve replication of similar technologies, may cater well to template-type appraisal approaches, with narrowly focused technical assessments coupled with fairly simple systems for rating financial risk. Focus on large projects, with investment costs of US$1 million or more, typically will require detailed feasibility studies, perhaps some customized technical design work, and greater technical sophistication. Large projects that require customized system-optimization approaches and in-depth knowledge of specific process technology may be the most demanding of technical expertise.

In theory, selection of target markets might best be accomplished following macro-level review of the overall energy efficiency market, thereby identifying the specific areas with the greatest potential for energy savings and commercial profit. In practice, however, market targeting often follows the perspective of the specific program developers and their institutional strengths, more than anything else. Specific public agencies may have specific goals, defined as part of politically determined national programs, which drive the direction of their program interests. Similarly, international donors may have particular niches in mind. Financiers often like to target groups where they want to develop customers or new business lines. For example, some banks may wish to focus on large industrial projects, with a view to expanding medium-term lending relationships with the large enterprises involved, whereas other banks may wish to expand small-scale business customer bases through promotion of cost-saving projects that are lucrative to these customers. Yet other financiers may wish to expand relationships with city governments or strong public utility customers, and hence may be especially interested in energy conservation projects in public utilities or other facilities. In addition, all project sponsors tend to have strengths and weaknesses in terms of institutional relationships at local levels, with technical groups, and with the financial community, which must be considered for a practical targeting of market priorities. Often the best solution for immediate impact is when new investment delivery programs for energy efficiency can build on other programs of sponsors that are already functioning for specific market segments. Examples may include existing specific targeted loan programs of banks, or specific technology promotion programs of public agencies.

For the energy efficiency program designer, then, the most practical approach often is a blend of review of areas with greatest potential with a review of the basic institutional options for delivery, and how the most traction may be gained by building upon institutional setups that already exist.

Market Outreach

Once initial markets are determined, outreach to end users to identify potential projects is an obvious next step. Evidence of a credible project pipeline is usually needed early on to convince potential program sponsors and financiers to support a new program effort. Once a program is up and running, institutional arrangements need to be in place to identify increasing numbers of projects on a sustainable basis.

Two points from the experience of various energy efficiency financing programs are especially relevant. First, initial market development, program outreach, and project identification efforts usually are a key first stage of necessary partnership between financiers and technical experts for energy efficiency investment project delivery. Few financiers can even begin to launch a new energy efficiency program without creating alliances for undertaking the necessary detailed technical review work. Even though they know that they will need the support of financiers for any program, energy efficiency practitioners and technical experts may try to undertake initial market development and project identification efforts themselves, with thoughts that they know what financiers will most likely require. There are cases where this has worked to some extent, but there also are ample cases where it has failed. Speculation about what will interest financiers (without sufficiently involving them) may often be proven wrong. Clients for prospective projects also may not take initiatives seriously unless they see the financiers specifically behind them. Thus, it is usually best to consider the characteristics and options for solid institutional partnerships between financiers and technical experts as early as possible, keeping in mind that costs and efforts in launching these new relationships properly will reap benefits for smoother program implementation later.

Second, nothing is as valuable in market outreach for a specific program as examples of projects using that specific program model that have successfully made money for the parties involved. Examples of successful cases provide clarity to the market as to specifics of the products being offered, and greater confidence to all parties. In launching a new program, therefore, it is often very helpful to focus on closure of just a few projects, as early as possible, under some type of pilot scheme—both to provide a basis for serious market outreach efforts, and to begin to test out the critical institutional partnerships.

In some cases, new programs may aim to create completely new markets. One example given in Part II is the creation of energy performance contracting markets in the United States and Canada by the development of new government programs (and necessary revision in public sector procurement regulations) to encourage energy efficiency investment in public facilities, such as schools or government office buildings (Case Study 8). Another example is the development of a market for energy efficiency investment in residential housing blocks in Lithuania, using new home owner associations (Case Study 6).

Such efforts face serious challenges, as they combine difficult policy and institutional development requirements. However, as these two cases show, large benefits can be realized over the longer term.

Once some clarity can be provided to prospective clients as to the specific project products being offered, perhaps with some successfully completed examples, market outreach usually relies upon the channels of the financiers, their technical partners, or other program sponsors. Trade associations, government agencies, business groups, or NGOs may also be enlisted to help expand knowledge of a program to prospective clients.

PROJECT DEVELOPMENT: IDENTIFYING, CLARIFYING, AND ALLOCATING RISK

Means to Identify Potential Projects

Specific candidate projects for financing through an energy efficiency program may be identified in a variety of ways. Projects may be identified by undertaking energy audits of facilities, which may be supported through some type of public program, or may be initiated by end users themselves. Energy audits are often done in two stages—the first being an initial, relatively simple audit, and the second being an "investment grade audit" that develops most of the basic data needed for credible, detailed project design. Solid energy audits provide the advantage of a fairly comprehensive review, usually with the objectivity of an outsider, of the most cost-effective and promising areas for energy efficiency improvement throughout a facility, and then confidence that the projects selected may be those best suited for the given facility. But audits are in some ways projects themselves, requiring agreement between parties and allocation of costs. Energy audits are by no means the only way to identify projects. Enterprises with strong technical capacity may identify projects themselves—especially medium-size and large industries and some public utilities. Third parties also may be another project identification channel. Equipment suppliers or some ESCOs may be in the business of marketing predefined basic solutions targeted towards specific equipment applications for specific types of consumers. Some examples include HVAC innovations, variable speed motor controls, boiler renovation projects, or waste heat or gas utilization solutions. The essential point at project identification stage is to gain confidence among the project stakeholders on the project's basic technical and financial viability. Further technical design, clarification of risks, and allocation of those risks among the various parties can then follow.

Detailed Project Design
Technical feasibility and design requirements for energy efficiency projects vary dramatically. Whereas some projects may require many months of engineering work to develop customized solutions for complex process technology improvements, design work for other projects may involve little more than checking technical and economic parameters and site characteristics for application of a predetermined solution, such as variable speed motor drives.

Many energy efficiency project feasibility or design reports focus almost exclusively on technical design and performance planning. However, to be financed, project assessments need to identify systematically all of the factors that may impact the projected energy cost savings cash stream, and assess the risks involved in each. This is critical if financiers are expected to consider the cost savings benefits of the proposed project accurately. Key risks to be considered are as follows:

(i) **Technical performance.** Satisfactory technical performance is the most important aspect of the energy efficiency investment. The project must perform and deliver the savings as designed. Confidence needs to be generated that risks have been minimized of the installed energy efficiency equipment experiencing total failure, or significantly lower performance compared to initial estimates. Risks linked to a specific piece of equipment (such as performance of a boiler or motor) are typically covered by a manufacturer/supplier warranty. Risks of lower-than-estimated overall energy savings from customized or integrated projects are usually borne by the client enterprise, the contractor, or project designer (for example, under a guaranteed-savings arrangement, see chapter 7), or some combination.

(ii) **Project implementation plan.** Project construction needs to be planned to minimize disruption in the production or use of the client's facilities. This typically is a key concern of clients, who stand to lose heavily if manufacturing lines are taken out of production for excessive periods or facilities cannot be accessed normally. However, delays in the procurement of contractors, construction, supply of equipment, testing, and so forth occur frequently. These issues are particularly important in larger projects that may involve a number of vendors or suppliers. Hence, implementation planning to minimize disruption, and up-front assignment of responsibilities and possibly penalties, is important.

(iii) **Capacity factor.** In the case of energy efficiency projects, the number of hours of use of the underlying equipment/processes each year is another variable that impacts energy savings. A variety of factors may be involved, such as weather in the case of HVAC systems in buildings, changing market conditions and consequent changes in production levels in the case of industries, and so forth. Such issues normally are addressed as part of defining the project baseline but still exist as a risk at the time of project implementation. This risk is normally carried by the end user, but can be shared or capped as specified in an energy service agreement.

(iv) **Quality of inputs and technology operation.** Another set of risks is linked to the quality of inputs (for example, poor quality of electricity supply or coal may result in poor energy efficiency performance). Normally, such risks would be outside the responsibility of any contractor or ESCO and would need to be shouldered by the end user. Similarly, lack of familiarity with new technologies and low capacity of the operational staff could result in suboptimal performance of the energy efficiency equipment; this risk needs to be considered and mitigated through training.

(v) **Energy price variations and attendant risks.** Even if the technical performance is at acceptable levels and operations are smooth, energy price variations could change the attractiveness of the project. It would be difficult to assign such a risk fully to an energy efficiency contractor, and therefore the end user typically assumes most of this risk. An explicit exception, however, is the chauffage model, where a third-party project developer essentially takes over complete responsibility for the provision to the client of an agreed set of energy services over a long term (see Case Studies 11 and 12).

Unless the end user is financing the proposed project with his/her own funds, both the end user and the financier will need to agree with the project designs and risk allocations. Their overall interests are in theory similar—that is, they need to be assured that the project net positive cash flows will be generated in sufficient levels and on time. However, their perspectives often do vary. If the financier is providing a loan backed by the client's balance sheet, the financier is likely to be most interested in the client's corporate financial standing and basic collateral and repayment guarantee arrangements. However, financier concerns about project details may be greater if the project is large enough relative to the corporation's overall operations (which could negatively impact corporate financial standing if there are problems) or if financing packages and repayment assurances are more narrowly constructed around the specific project. On his/her side, too, the end user is likely to bring particular perspectives. Typically, the end user is interested in how the project may have a favorable impact over the longer term, and not just its ability to generate resources for loan repayment. End users can be expected to be particularly interested in how the project fits into overall improvement plans, and may help improve product or service quality, or spark other efficiency gains, as well as reduce energy costs.

A key point is that *both* the end user and the financier must trust the technical assessment and be confident that it meets their quality needs for the investment to proceed. This means that the financier will most likely require a review and evaluation independent from that prepared or commissioned by the end user enterprise. If a technical contractor used by an enterprise for project development has a strong reputation, this may reduce requirements of the financier to delve into details. However, some type of independent review of project specifics by the financier or its agent is still likely to be necessary.

For evaluation of project design and risk assessment, therefore, both the end user (or project host) and the financier must mobilize some level of technical skills to independently evaluate the potential project from their own perspective.

INSTITUTIONAL CAPACITIES FOR ENERGY EFFICIENCY PROJECT DEVELOPMENT

A wide variety of institutional mechanisms have been utilized globally for completing project development and technical assessment work on energy efficiency projects. These frameworks vary due, in part, to the following factors:

(i) The levels and extent of energy efficiency technical evaluation and project development capacities that exist in the country. This is determined by the level of engineering expertise, the level of energy audit awareness and experience, the level of energy efficiency project development experience, capacity of training facilities available, and the number of qualified energy auditors operating (sometimes certified and licensed), and their understanding of the financial issues related to project formulation and design.

(ii) How technical capacities are organized within each society. Although technical capacity may exist, it may not be easily accessible by different institutions, depending on prevailing relationships between contractors and contractees. Often, a given financier or enterprise may not know where they can go for trusted, qualified assistance, especially if the expertise is not available within their normal circle of contacts.

(iii) How the energy efficiency policy and legal environment has evolved in each country. Energy performance contracting may be legally "unknown" in a given country, and pose excessive risks related to contract enforcement. Another example is that some countries have very involved systems of certifying energy auditors and energy efficiency practitioners such as ESCOs, in an effort to provide comfort to prospective clients, and others may not.

Donor assistance for energy efficiency initiatives may influence both the level of project development and technical capacity and its organization and areas of coverage in specific countries. Typically, training and other capacity-building activities are initiated (in some cases in a fairly aggressive manner) to increase the technical capacity and project development skills of practitioners. In some cases, specific assistance is given on a grant or a contingent grant basis to prepare a pipeline of energy efficiency projects. The type of assistance can include providing a "finder's fee" to consultants and other agents who bring viable projects for consideration; initiating twinning arrangements with particular customer segments such as industry associations to generate energy efficiency proposals; and providing broad-based training to ESCOs and energy auditors.

In-Country Technical Capacity
China, India, and Brazil all have a high level of technical capacity to undertake energy efficiency work. There are universities or other educational institutions with departments or programs focused on industrial productivity and competitiveness, technology upgrading, energy management, and energy efficiency. Furthermore, there are numerous professional short-term training programs available in each of these countries. Many government and donor programs have provided support for energy audits, spurring the development of the energy auditor industry. This energy efficiency capacity is built on a fundamentally sound technical education system in the three countries, which produces large pools of qualified engineers and technicians.

The main technical challenge for energy efficiency in China, India, Brazil, and other developing countries concerns the organization of these technical skills in the energy efficiency institutional context, and how effectively these skills are being accessed to provide needed technical assessments to enterprises and financiers.

China has a tradition of several decades of publicly sponsored energy efficiency efforts. Energy efficiency centers and energy auditing groups were established in the 1980s under the planned economy to help identify energy savings opportunities and complete technical design work. With over 200 local or sectorally focused energy technology service centers, thousands of staff have been trained and deployed. In addition, technical staff in the corporate world are often quite strong. Even so, the task of improving energy efficiency in China is so large that is has become a national priority to expand existing technical capacities, to provide yet broader and more sophisticated energy auditing and project identification and development services.

India also has a long history in the area of energy conservation and energy management. The first government efforts in the public sector started in the 1970s and were initially centered on productivity improvement and oil conservation. The "energy audit" tradition started in the 1980s and has steadily grown since then through the efforts of the government and numerous donor programs. Institutions such as the National Productivity Council, the Ministry of Power, universities, and NGOs (such as Tata Energy and Resources Institute) built the in-country technical capacity in the area of energy auditing, with a particular focus on SMEs, but also to some extent on the commercial sector. Government support came in the form of support to universities, technical training facilities and energy audit subsidies to consumers. In 2001, the government passed the Energy Conservation Act, which created the Bureau of Energy Efficiency (BEE) to institutionalize public energy efficiency services and provide leadership to the key players involved in the energy conservation movement. The Act calls for implementation of a certification and accreditation program for energy auditors, adoption of technical standards, and implementation of an energy manager training program that will improve in-house technical capacity for energy efficiency in industries.

In Brazil, technical capacity and project development experience in energy efficiency were mainly developed through the national electricity conservation program (PROCEL), established in 1985. PROCEL, managed by Eletrobras, the federal holding company in the electricity sector, operates by funding or cofunding energy efficiency projects such as R&D, education and training, testing, labeling, standards, demonstration, and others. The program works on both increasing end-use efficiency and reducing losses in electricity generation, transmission, and distribution systems—though the latter diminished greatly after the liberalizing reforms of the 1990s. PROCEL cooperates with state and local utilities, state agencies, private companies, universities, and research institutes. Donor support has played an important role in training of energy efficiency technical expertise. In the case of ESCOs this support was largely channeled through ABESCO, the Brazil association of energy service companies.

Although there is strong technical assessment capacity in China, India, and Brazil, technical experts often may not be well linked to the business environment, and the practical needs of end users and financial institutions. Challenges remain to build the investment project development and risk assessment skills needed to expand commercial energy efficiency investment.

In many smaller countries, adequate local technical capacity may not be available. Use of foreign consultants may be useful if well targeted, and focused on helping local capacities develop. Such reliance, however, also can be risky and unsustainable. Foreign consultants may have difficulty adapting to the local working culture. Costs typically are high, and the outside consultants may not be available over sustained periods of time. In such situations, utilizing simpler project designs and regional expertise can be helpful. For example, countries such as Thailand, Japan, and India in Asia, and Brazil, Argentina, and Mexico in Latin America, often have served as regional hubs for energy efficiency project development, technical expertise, and financial advisory skills.

Combining Technical and Financial Risk Assessment Skills

A combination of financial and technical skills is necessary to successfully deliver energy efficiency projects—yet institutional combination of these skills is rarely available. Usually, financiers must contract for technical expertise to combine an independent technical evaluation with financial assessment. Among the technology assessment and project design companies that may work for end users, true financial expertise and skill in analyzing project risks outside of purely technical risks is typically thin. Development of capacity in these areas is especially needed.

In general, one of the great challenges in delivering energy efficiency investment is linking up-front project identification and development work to actual project implementation, particularly when audits are subsidized by governments or donors. Numerous project development activities were undertaken in previous years without accounting for nontechnical issues, leading to lists of unfinanced projects and wasted resources. Well-designed technical assistance programs can be targeted at generating a healthy pipeline of viable projects that can then be funded through specific loan funds (as in India, Romania, and Lithuania), or a guarantee facility (as in Hungary and China). However, effective linkage between project development efforts and financing programs is essential.

MAKING CHOICES ABOUT OUTSOURCING

Both financiers and end users must decide to what extent technical assessment work should be outsourced. As discussed in chapter 4, decisions on whether to develop and use internal expertise versus using contracted external expertise basically involve weighing the various benefits and costs involved for each alternative. These will vary according to industry/sector and according to the particular institutional environment posed by the country's legal system and local ways of doing business. In practice, as we have discussed in previous chapters, there is no universally right or wrong approach, apart from the general guidance given by the model outlined in chapter 4.

Outsourcing by End Users

Among end users, big differences typically exist between large industrial clients and small industry, building owners and public sector energy users, in their internal capabilities and needs and desires for outsourcing of technical assessment work.

For many large, energy-intensive, industrial companies, energy management is a factor in remaining competitive, and these companies often pay significant attention to identifying and implementing means to reduce energy costs. Specialized internal energy management units may be justified, which typically can undertake energy efficiency project development work. Even in these cases, however, firms often hire outside expertise to help introduce the latest technology or help tackle complex design issues. Because coordination between internal and external experts must be effective, however, firms will tend to contract experts from institutions that are well known to them.

In smaller enterprises and among firms where energy costs are not a particularly large portion of overall operating costs, employment of internal energy management expertise is usually not justified. Outside expertise of some type is then usually necessary to develop an energy efficiency project. Firms often may not have prior relationships with groups competent in energy efficiency work; indeed, identification of whom to trust may often be a big first challenge.

Outsourcing by Financiers

Among financiers, there are a few cases where strong internal capacities for technical assessments are developed as part of an effort to create a one-stop shop for clients—for example, in specialized lending institutions such as IREDA in India (Case Study 4) or the Romanian Energy Efficiency Fund (Case Study 3). But even here, some outsourcing may occur for specialized needs. For the most part, however, financiers do not wish to build up their own internal capacity to do such specialized technical assessment work in house. Instead they generally prefer to rely on contractors, but with the possibility of a check from their own appraisal department or other "close to home" experts. Meanwhile, it would be challenging for outsiders to independently develop and tailor an external delivery option for technical assessments for banks or other financiers for an energy efficiency financing program; this is because financiers will want opinions that they can trust (see the suggestions at the end of this chapter regarding industry groups and certification standards). Hence it is usually best for financiers to make their own arrangements for accessing this service. In the case of the emerging Brazil PROESCO guarantee facility (as described in the Brazil country report), for example, a panel of independent evaluators is being proposed to assist the BNDES and the participating commercial banks in the appraisal of project proposals.

Outsourcing Options
Some of the options available for obtaining outsourced expertise for project development and/or assessment include the following:

- **Equipment vendors.** In some cases, the end user may have a good idea of the energy savings intervention required on his/her premises, particularly if a single type of equipment is concerned, such as boilers, capacitors, or well-defined efficient lighting systems or building energy management systems. In such instances, enterprises may often use the services of equipment vendors who normally provide such technical assessment services to buyers as part of their marketing process. Of course, such technical assessments usually will result in recommendations to install at least some of that vendor's own products. An advantage is that some vendors specialize in developing client-specific solutions and rest their reputations on the quality of these solutions. Further, costs related to technical assessments completed by vendors working with client staff should be reasonable. However, the technical assessment may be limited by the vendor's perspective of energy efficiency and the vendor's desires to sell its products. Again, relying upon vendors for technical assessments is generally best done when the client has in-house expertise that is able to work with and manage the vendor's technical recommendations to meet the client's needs.

- **Private energy auditors.** This is the most common delivery option available to end users to outsource energy efficiency technical assessments and project development. Energy auditors offer different levels of service such as "walk-through" audits or "detailed/investment grade" audits. Energy auditors might simply be individual consultants who have training and experience in conducting audits, or the audits might come from larger, private consulting firms. In either case, audits often include a thermal and an electricity audit component to cover all energy end-uses on the client premises. The advantage of utilizing energy auditors is that they bring a fresh and independent perspective of the facility and its operation and can help the end user to prioritize energy savings options. However, independent auditors may have a limited understanding of the client's production process, increasing the danger that the auditors' recommendations could be difficult to implement. Another downside of energy auditors is that they may be "study oriented"— better at writing reports than at designing implementable solutions—and may not be able to meet the assessment requirements in a manner that meets the needs of the end users and the financiers. In India, SBI relies on "empaneled" auditors from the BEE list of accredited auditors to address issues of bank comfort in auditor technical capacity.

- **Project appraisal companies.** In some countries, specialized project appraisal companies are available to undertake technical assessment work. Such companies specialize in providing neutral technical appraisal services, and financiers as well as end users can seek their guidance while making investment decisions. In China, project appraisal companies are often hired by either large end users or financiers to provide independent project assessments.

- **Universities/research institutes.** Technical assessment services are often available through universities, research institutions, public agencies, and design institutes. Typically, their technical assessment services are reliable and inexpensive, if sometimes somewhat academic in nature and less anchored in operations. However, such a delivery mechanism can be useful in specific cases where unbiased technical assessment is needed.

- **ESCOs.** ESCOs are emerging around the world as providers of a range of services related to energy efficiency technical assessment and project development. These could range from professional energy auditing services to single-window service for project development and implementation. In order to develop their business, ESCOs actively market their services to clients and develop pipelines of potential projects. This mechanism largely works within the private sector based on contractual agreements with clients and/or financiers. Although they are private companies themselves, ESCOs often are allowed to benefit from training and other nonfinancial assistance through donor and government-run programs designed to enhance energy efficiency delivery capacity in the country. Typically, ESCOs work through energy performance contracts (EPCs) that specify *what* is to be achieved rather than *how* it is to be achieved. ESCOs are increasingly used by government agencies to outsource the provision of energy efficiency services to public sector clients (see Case Study 8, for example), as well as by private sector clients in both commercial and industrial sectors (see Case Study 7). Although ESCOs mostly are capable of developing technical proposals even in developing countries, the development of "bankable" proposals that can access financing is still a challenge.

In the end, outsourcing decisions are determined by key issues such as trust in the outside service provider, the current state of evolving contracting and enforcement institutions, cost considerations, and the perception of how much value might be added by building specialized capacity within the enterprise. This is equally true for end users and financiers.

SOME OPTIONS TO MINIMIZE TRANSACTION COSTS

There are multiple incentives for the various stakeholders in the energy efficiency industry to reduce the various transaction costs associated with the project development cycle.[1] The initial costs for energy efficiency market development and for individual project development are high-risk outlays, and are typically unrecoverable if the identified project or program is not implemented for whatever reason. Thus, as outlined above, donor and government programs often have important roles to play in helping defray some of the risk that initially would have to be borne by individual market actors, if they decided to proceed on their own.[2]

There is a high incentive both for clients and for energy efficiency project developers to minimize these up-front costs until certain basic questions regarding technical and financial feasibility of potential efficiency investments can be answered. Additionally, due to the relatively small size of many energy efficiency investments, high transaction costs can be a key impediment to development and implementation of energy efficiency projects on both the end user and the financier side. Some options for reducing these types of transaction costs at the firm and the financier levels include the following:

- **Selection and replication of standard projects and templates.** Simple project designs can help minimize transaction costs and reduce risks. In particular, for many of the technologies well proven in the market (such as efficient boilers, building energy management systems, efficient lighting systems, and capacitors), project design is relatively simple and the end users and financiers are reasonably well aware of the risks. This also holds true for certain industry-specific technologies and applications, such as waste heat recovery boilers or natural gas cooling systems with cogeneration (see Case Studies 11 and 12). In such cases, it is relatively easy to follow a template in the replication of design and appraisal of new projects, which often will significantly reduce transaction costs for standardized (or "commoditized") projects. Often equipment vendors offer such template-oriented energy efficiency project schemes that can be customized to fit the end user needs.

- **Clustering and specialization in project identification and development.** The cluster approach represents a method to provide specialized technical support and outreach to groups of smaller enterprises and follow-up loan provision under a standardized approach to allow for significant replication and reduction in transaction costs. Typically, there is a high fallout rate in the early stages of project development, which may translate to high transaction costs per project. One way to reduce such costs is by targeting a group of end users with similar characteristics (in other words, a cluster). Cluster lending has been implemented in India with either a specific sector/technology focus, or with geographically grouped clusters that include several industrial categories but concentrate on a few technical interventions to reduce assessment and appraisal costs. (A more detailed discussion of this approach is presented in Case Study 5.)

There are a few simple "public goods" kinds of steps that can be taken by governments or by industry groups, including the following:

- **Use of strong and credible expertise.** Financiers and end users place a high value on having agents with good reputations from past experience provide technical expertise. Developing a cadre of agents certified by the government or another apex agency provides a certain level of comfort to the end users and financiers and can reduce costs associated with verification of claimed savings from proposed projects. Being part of a recognized professional organization such as an ESCO association can also provide credibility to the firm or individual offering to undertake the technical assessment. The ESCO associations in Brazil and China play this role, although to varying extents, and they also are considering design of certification procedures for their members.

- **Use of strategic partners.** In some cases, the service provider is a subsidiary of an industrial group. In particular, the institutional environment in many developing countries makes relationship-based contracting more common than more legalistic, arms-length approaches followed in the Western countries. As such, the end users representing group companies have a high level of comfort seeking technical assessment service from another company related to the group.[3] The charges for the services tend to be reasonable on the whole, avoiding both the ex ante transaction costs of developing formal contracts and the ex post transaction costs of enforcing and adjudicating them. Governments and industry leaders can play valuable roles in stimulating development of such industry groupings.

NOTES

1. Perhaps the sole exception is those lawyers whose income appears as a transaction cost borne by clients and financiers of energy efficiency projects. In some cases these lawyers would be losers in the "public goods" approach to transaction cost reduction discussed in the following footnote.
2. We have previously suggested that second and third entrants into the energy efficiency finance market, for example, get the benefit of the first entrant's risk-bearing and market-development investments without sharing in these up-front costs. This provides a "public goods" argument favoring government and donor assistance with upfront transaction costs affecting more than one potential market participant or entrant. It is important, of course, that the aggregated net benefits enjoyed by all of these actors add up to more than the up-front transaction costs borne by governments or donors for the related project or program.
3. Fernandez and Kraay (2005), for example, document the more general tendency for companies in India to place greater reliance on membership organizations in states having weaker formal contracting institutions (judicial systems).

CHAPTER 6

DELIVERY OF FINANCING

WHAT NEEDS TO BE DELIVERED?

On the financing side of energy efficiency projects, the desired result is straightforward: provision of the financial resources to complete project investments. This involves a set of transactions to convert a flow of energy savings into a capitalized energy efficiency investment.

Financing may be found from any of a variety of sources, including internally generated funds of enterprises themselves; informal arrangements for loan or equity financing from shareholders or other financiers related to enterprises; formal loans from financial institutions; or, occasionally, various types of equity injection. For sustainable and sizable channels of financing, however, the local banking sector is ultimately the key in almost every country. Where the local banking sector is too weak, immature, or simply uninterested, any of a variety of arrangements may be worthwhile to provide initial energy efficiency project financing or begin to introduce the energy efficiency lending business into the market. These may involve other public institutions in a variety of ways. Ultimately, however, effective long-term solutions are bound to require large-scale, well-constructed involvement of the local banking system.

Discussion in this chapter begins, therefore, with a review of the institutional environment of local banking sectors.

THE INSTITUTIONAL ENVIRONMENT FOR FINANCING: THE DIFFERENT WORLDS OF BANKERS AND ENERGY EFFICIENCY PROJECT PROMOTERS

One of the most valuable aspects of the UNF Three Country Energy Efficiency Project was the development of sustained dialogue between representatives of the energy efficiency and banking industries in each of the three countries. The difference in perspectives of these two groups was striking. In the earliest meetings and cross exchanges, there was a clear lack of mutual understanding. Through discussion of practical issues, concrete proposals, and experiences gained, however, ideas began to emerge on how energy efficiency financing could be developed through existing banking structures. Some of the bankers began to identify new business opportunities in the energy efficiency sector, and many of the energy efficiency practitioners gained a far better understanding of the needs and perspectives of the bankers. New energy efficiency financing programs involving banks participating in the project were identified, and those new programs now are either under active preparation or early implementation. But major efforts were required to overcome the lack of mutual understanding at the beginning.

The View from Energy Efficiency Promoters

Energy efficiency project promoters may include ESCOs, industrial or other end-user representatives, equipment and technology providers, or a variety of different types of energy efficiency experts. These experts typically focus on the technical, economic, and financial merits of specific projects. Armed with feasibility studies demonstrating low technical risk and high financial returns from energy cost savings, these project promoters often find their efforts to obtain loan financing from local banks frustrating.

Ultimately, energy efficiency project promoters are especially interested in project-based financing, which recognizes the key financial benefit streams of energy cost savings derived from energy efficiency projects. Project promoters may understand and accept the need for banks to obtain additional assurances that the technologies proposed will generate the energy savings forecast, through additional third-party appraisal and/or contractual guarantees. But the nature of the energy efficiency investment business generates desires for the cost savings revenue stream of projects to be clearly defined and then recognized somehow, at least partially, as a form of loan security. Fixed asset values from energy efficiency investments are usually (but not always) substantially less than loan amounts, especially where projects involve a series of relatively small measures and significant soft costs. On the other hand, cash benefit streams from the cost savings of most energy efficiency projects are typically quick and very robust. Hence the desire to somehow include these factors into the financing security equations of the banks.

The View from Local Banks

The views of the different banks vary considerably, since country banking environments vary substantially, as do individual bank cultures and business focus. Based especially on the inputs of the banks participating in the Three Country Energy Efficiency Project, but also considering many experiences and discussions in other countries, the following observations can be made about how banks tend to view energy efficiency investment projects:

- **Energy efficiency projects represent a relatively small, niche business.** For most banks, there is nothing special about energy efficiency, and, by itself, this is a relatively narrow market. Commercially oriented banks usually like to be perceived as "good citizens," but the public benefits of energy efficiency are largely irrelevant for their business.

- **Cost savings project finance is nonconventional.** The mainstay in lending to industrial and commercial enterprises is working capital finance—lending to sustain the working capital cycle of enterprises. Lending against the balance sheet of enterprises is strongly preferred. Business lines may also include project finance to build new production capacity, but this is usually concentrated on large projects. In Brazil, China, and India, lending for projects to improve business efficiency and increase productivity is uncommon.

- **Banks lack knowledge of energy efficiency technology.** Most bankers would consider gaining specialized knowledge of energy efficiency technology to be outside the scope of their operational interest.

- **Existing procedural frameworks are important.** Although systems vary substantially between banks, loan generation, risk analysis, appraisal, and supervision procedures are well established, and built for existing mainstream business lines. To be operationalized effectively, new lines of business need to fit into existing systems. For large banks, a key aspect of this also is division of responsibilities and expertise, as well as working relations, among central, regional, and local branch offices.

- **Customer relations are important.** New product lines may be of strategic interest to banks as a new service to offer key existing customers or as a means to attract new customers, depending on the type of product and the bank's marketing plans. Bank business strategies in this regard may dictate where different banks are interested in developing which types of loans.

- **Transaction costs for small projects are often a key issue.** Most energy efficiency projects are considered small loans. Interest in small loans varies depending upon the bank and the bank level. To be undertaken efficiently, however, transaction costs need to be minimized, usually through use of easily replicable loan products and appraisal methods.

An additional, major set of issues in many developing countries and emerging market economies arises from the fact that local banking sectors are often less developed or mature when compared to banking sectors in industrialized countries, or constrained by uncompleted banking policy reforms or any number of other banking framework distortions.

DEALING WITH BANKING SYSTEMS IN TRANSITION OR UNDER DEVELOPMENT

In some countries, the local banking sector may be close to dysfunctional, the policy and operating environment severely distorted, or a major uncertain transitory reform process may be underway. This can make proper financial intermediation of energy efficiency investment projects through the banking sector close to impossible. The implementation record of financial intermediation projects within banks attempted under such conditions is very poor.[1] With these conditions, energy efficiency proponents may face a difficult choice of attempting to shoulder the increased difficulties and risks of working in an immature banking sector, developing a solution more independent from the banks, or foregoing efforts to promote energy efficiency investments until the local financial sector is in better shape. A decision to proceed, especially with an independent or semi-independent approach, may or may not be warranted, but the high risks and needs for intensive efforts during implementation, including flexibility for major midcourse corrections as conditions evolve, should be recognized up front.[2]

The deep-seated issues in the banking sectors of both Brazil and China provide additional challenges for energy efficiency investment financing.[3] In **Brazil**, prospective borrowers face an exceptionally tight credit environment. Bank lending to government comprises a very large and profitable portion of local bank business. In addition, risk-averse attitudes remain from a history that included major losses in enterprise credit operations during the high-inflation eras of the 1980s and 1990s. Real interest rates to medium and especially small enterprises are extraordinarily high, collateral requirements are extremely strict, and available loan tenures are short. Loans with resources of the BNDES—at moderate interest rates and longer tenures—are possible, but guarantee requirements are severe. Banks are generally quite profitable lending at these terms, and there is little incentive for them to take risks, especially in new niche markets. In 2003, credit to the private sector in Brazil amounted to only about 25 percent of GDP, compared to about 62 percent in Chile and some 115 percent in the Euro area. Brazil's banking community is quite sophisticated, and banking experts are highly skilled. Competition in the banking market also is increasing and some banks are giving more attention to investments to increase firms' competitiveness. However, although a more "borrower friendly" environment may be approaching, it does not exist as yet.

China's banking sector is just beginning to emerge from a decades-long process of reform from a state-administered credit system to a market-based, competitive banking system. Current interest rates are attractively low to borrowers, and are still profitable to banks. However, as in Brazil, banks currently have little incentive to take risk, although for different reasons. Although interest rate ceilings were basically removed in 2005, rates have long been set by government regulators within a narrow band. With banking returns roughly similar across the risk-profile spectrum, there has been little incentive to take on anything but minimal risk. In addition, the older, state-owned banks still have high ratios of nonperforming loans stemming in large part from the previous era, adding to risk-averse policies, including from bank regulators. Potential energy efficiency project borrowers, therefore, face exceptionally tight and very inflexible requirements for collateral and counter guarantees, and reticence to move from short-tenure loans, or to try new, innovative approaches to risk mitigation, for which little additional bank return is seen, at least over the short term.

In contrast to China and Brazil, the **Indian** commercial banking sector offers a healthy degree of competition amongst the numerous public and private banks. Liquidity is high, rates are attractive, and there are numerous alternative products available to borrowers. In addition to the traditional working capital and term loan products for entrepreneurs, commercial banks have started to offer structured products to meet the specific needs of certain classes of borrowers. However, public sector banks are still the largest credit providers in India. These types of banks have traditionally served social and other noneconomic objectives of the government and are also the major holders of nonperforming assets. Recently, the Reserve Bank of India (India's Central Bank) has strengthened regulatory and supervisory norms to induce greater accountability. Even though small-scale industries are included in the "priority sector" listing for bank lending purposes and receive considerable fiscal and other incentives, small firms still have very limited access to commercial bank credit and rely mostly on alternative financing channels. This goes hand in hand with entrepreneurs and investors relying more on informal governance mechanisms, based on reputation, trust, and relationships, rather than formal mechanisms such as courts to resolve disputes, overcome corruption, and finance corporate growth.[4]

INSTITUTIONAL OPTIONS FOR DELIVERY OF FINANCING

In principle, formal mechanisms to deliver financing for energy efficiency investments include the following:

- **End-user self financing.** Host enterprises usually contribute a significant share of the investment financing in most energy efficiency projects from their own retained earnings. In addition, small or even medium-size energy efficiency projects in major enterprises may be financed internally.

- **Local banks.** Commercial or policy banks operating in each country are the main source of external loan finance for a variety of projects.

- **Nonbank financial institutions (NBFIs).**[5] Special financial institutions exist in many countries, usually relying primarily on public sector support, which provide lending for energy efficiency investments, as a special priority of public interest. India's IREDA and China's Energy Conservation Investment Corporation (CECIC) are two examples.

- **Leasing companies.** Depending upon local tax and leasing laws, leasing companies can provide an important vehicle for commercial debt finance for energy efficiency projects. Many banks have leasing subsidiaries. In other cases, stand-alone leasing companies are often more aggressive than banks in probing new market segments. Leasing company partnerships with energy efficiency equipment companies and vendors can provide an important vehicle for marketing energy efficient technology with a financing solution. Such partnerships can provide important secondary markets and recourse vehicles to support credit structuring.

- **Multilateral development banks (MDBs).** The World Bank, including IFC, and regional development banks may provide direct financing to large end users, especially energy utilities. With a few exceptions, however, energy efficiency investment financing outside of the energy utilities from these institutions needs to be channeled through financial intermediaries (such as banks or NBFIs) to be efficient.

- **Others.** ESCOs may be a source of financing to end users, but ESCOs themselves then require financing, through one of the options above. Energy utilities may occasionally provide financing to end users, sometimes with repayment factored into utility bills. In a few cases, major energy efficiency projects may be financed with equity capital through special-purpose companies. Suppliers' credits for equipment purchase in energy efficiency projects may be important, especially in industry. Capital market instruments, such as bond issues, have been used in more mature markets, such as for public building energy efficiency programs in the United States (see Case Study 8). Finally, financing from enterprise shareholders or through informal means is often a substantial, if difficult to document, source of capital for enterprises in developing countries, which may be important in energy efficiency investment.

In principle, **equity finance** for energy efficiency projects or service providers such as ESCOs constitutes another source of financing. After all, to accelerate their growth, ESCOs need a growing equity base to accommodate more debt financing. However, investigations within the framework of the Three Country Energy Efficiency Project found the potential for equity financing from venture capital and private equity funds in this industry to be quite low.[6] The extent of equity financing and the potential are slightly different in each of the three countries, but there is agreement on the factors for the low potential:

- Few ESCOs so far have been able to attract equity injections since ESCOs are typically small and project-based. The small scale usually precludes any due diligence by equity investors since the costs are fairly high and need to be justified by a substantial size of the investment.
- The failure of most ESCO managements to clearly articulate business plans that establish a clear vision and strategy for growth and the lack of sufficiently transparent financial and accounting systems contribute to the lack of interest from investors.
- Standard documentation for performance contracts, monitoring and verification protocols, and financial instruments and structures is not yet available in a customized form in many countries. This may reinforce the perception of equity investors that scaling up the ESCO business is not yet possible.
- Equity providers are looking for high returns that are not obtainable without the leverage available from additional debt financing, which many ESCOs have problems obtaining.
- Exit strategies such as initial public offerings (IPOs) or strategic investors are not obvious in the energy efficiency/ESCO industry, which leaves potential investors with a risk of not being able to divest themselves.

To provide sustainable and sizable flows of finance for local energy efficiency investments it is clear that ultimately the key focus should be placed on local commercial banks. However, where the local banking sector is not mature or is weak for any of a variety of reasons, it may be difficult if not impossible to build energy efficiency financing programs with local banks. Moreover, even if local banks are healthy and sophisticated, it may be difficult to engage them in energy efficiency work immediately and directly, without some prior market development work. Often, therefore, other options have been adopted, including use of policy banks, development of energy efficiency lending programs in NBFIs, use of credit enhancement schemes in association with the banks (such as guarantee funds), revolving fund schemes, or use of banks as agents to disburse for sets of previously appraised energy efficiency investments. Use of these options may or may not be warranted or ultimately successful, depending upon the objectives and circumstances. In many cases, use of these options is seen as a transitional measure—as a way to open the energy efficiency lending market, and hopefully develop large and sustainable financing flows from commercial banks. In these cases, use of these options is seen as a means toward the end result of increasing commercial bank involvement. Effort must be taken, however, to drive toward the foreseen end result, and not confuse the means with the ends. As stated previously, in the end, development of energy efficiency lending programs in local commercial banks has the best prospect for sustainable, large-scale results.

Advantages and disadvantages with a wider range of financing delivery mechanisms, and lessons learned from experience, especially in Brazil, India, and China, are discussed in the next chapter, where different institutional mechanisms combining delivery of financing and technical requirements are evaluated. However, one set of issues that financiers must deal with in virtually all of the delivery constructs involves the structuring of financing products to match the needs of cost-saving energy efficiency projects. In essence, what are some of the options to meet the needs expressed by the energy efficiency promoters in financing products, which were mentioned before? This question is explored below.

DEALING WITH REPAYMENT ISSUES IN ENERGY EFFICIENCY PROJECTS AND NEW FINANCING PRODUCT DEVELOPMENT

First, it is important to emphasize that many energy efficiency projects can be attractively financed using existing bank loan products, with no special adjustments or new financial product development. If the financier is convinced of the technical merits of the project and of its ability to generate the revenues forecast, it is often possible to use existing loan programs and financial products without any major adjustments. Banks may be able to introduce energy efficiency loans into lending relationships with existing customers quite easily. In the energy efficiency lending programs for SMEs developed by five major Indian banks (see Case Study 5) with support from this UNF Three Country Energy Efficiency Project, the main challenge has been, in addition to marketing this new product effectively to enterprises and project promoters, to make efficient arrangements for the technical assessment and appraisal part of the process. The development of special financing products was, in most cases, not required. SBI, for example, simply slotted the energy efficiency program into its existing Project Uptech financing program for SMEs, using the same credit review and loan terms and agreements as for other parts of that program.

In some cases, difficulties experienced by energy efficiency project promoters in obtaining financing are due to poorly prepared project proposals, lack of experience in applying for financing, or basic lack of understanding of the needs of financiers to mitigate lending risks. These failures, then, should not be translated into unsubstantiated demands for bankers to change their approaches, but instead should be translated into needs for the project developers to better understand basic bank financing requirements. This may, at times, require special technical assistance arrangements, where needs of project developers are legitimate. During the first two years of operation, many of the loan guarantees provided under the China ESCO Loan Guarantee Program (Case Study 1) were to ESCOs that had never borrowed money formally from any institution before. I&G, which operates the guarantee program, estimates that on average 2–4 months were required during early stages of the program to process loan guarantee applications from such companies, as major efforts are required from I&G to familiarize these new borrowers with the realities of the world of formal commercial credit.

There is no doubt that development of financial product modifications to match the characteristics of energy efficiency projects (or any of a variety of investment projects to improve business efficiency and productivity, for that matter) could help expand the market for such loans and could increase uptake of financially viable, yet unimplemented projects. In the opinion of this study team, however, development of new financing products should not be considered mandatory for a commercially viable energy efficiency lending program. Rather, development of new products to meet specific client needs in specific markets should be seen as a way to expand and strengthen such business.

The main, strategic focus for development of modified financing products for the energy efficiency market is to develop mechanisms that recognize and define the cost-reduction cash flow benefits of the projects, and use this flow of funds as a source of loan repayment and security. However, an objective to develop nonrecourse project financing products based solely on the cash flow from operating cost savings is overstated, and generally, unrealistic. The key is for financiers to increasingly recognize the characteristics of the cash stream generated by the projects financed, and to structure loans and repayment assurances to best take advantage of that. This is an art of developing enhancements and modifications, grafted onto existing primary loan products. It should be noted, however, that technical assessments need to properly identify, and mitigate to the extent possible, all of the risks associated with the delivery of the project cash flow, and not only technical performance.

A number of tools are briefly described below that have been used by financiers for energy efficiency projects in a variety of developing countries and emerging economies as mechanisms to partially mitigate repayment risks from borrowers, using in some way the generated project cash flows.

Matching loan repayment schedules to project cash flow. Loan repayment risks can be reduced by agreeing on more frequent loan repayment schedules (for example, more frequent than once per year) and repayment schedules that include both principal and interest payments (similar to mortgage payments). The cash flows of energy cost savings typically accrue fairly regularly over time, and assignment of these cash flows more regularly for loan repayment can help reduce risks that this cash will be used for something else or that loan repayment difficulties will arise when loan principal payments are due.

Use of escrow accounts. Adding further to the option of matching repayment schedules to project cash flow is the option of requiring hosts to make periodic payments from energy cost savings into escrow accounts, with account control assigned to lenders, to ensure loan repayment. Case Study 13 (India capacitor leasing) demonstrates use of this mechanism. A variation is for clients to deposit cash in an escrow account for payment of energy utility bills, at historic levels, and for these funds to be used for paying the utility and repaying energy efficiency loans. Use of the escrow account tool, however, may require resolution of various legal issues, including, for example, the legality of different arrangements for assignment of client cash flows to banks or their guarantors, and clarity on third-party creditor rights to such escrowed resources in the event of client bankruptcy.

Use of energy efficiency performance guarantees. Third-party entities such as ESCOs may provide guarantees, backed with financial resources, on the energy savings performance of energy efficiency investments, thereby reinforcing the principle that risk should be apportioned to parties best in position to manage that risk. Such guarantees can cover different parts of a range of risks associated with the realization of the energy cost savings cash flow, in addition to the technical performance of the project itself. However, a key issue is the scope and strength of the financial backing of such guarantees. A common problem faced by ESCOs is inadequate capitalization to actually stand behind any performance guarantee they seek to offer into the market.[7]

Use of ESCOs as project aggregators. Third parties, such as ESCOs, that provide some portion of project financing to clients, may act as project aggregators for financiers, which then can provide lines of credit or other types of loans against the financial strength and/or portfolio of the aggregator. This trend is developing with the stronger ESCOs in China (case studies 1, 7, and 11). Risks to financiers can be reduced by diversifying portfolios, by implicit or explicit recourse to accounts receivable from prior energy efficiency investments exhibiting stable repayment, and/or by recourse to other financial assets of the project aggregator or its shareholders.

Payments through utility bills. In some cases, especially where an energy utility acts as financier, repayment may be made through the utility billing system, allowing a steady, regular payment stream, and, at times, additional repayment protection. Case Study 10 (Sri Lanka DSM) provides an example. In many cases, however, governments or regulators look askance at addition of loan repayments into utility bills, preferring to keep the utility/customer contractual relationship implicit in utility billing simple and straightforward, and resisting, in particular, provisions allowing customer disconnection due to loan repayment default.

Build-own-operate or build-operate-transfer cogeneration chauffage. An attractive business model for industrial cogeneration projects is for project developers to build, own, and operate (with or without eventual asset ownership transfer to the client) assets within enterprises to generate electricity for the client, using waste gases or waste heat also generated by the client. The project developer is responsible for the cogeneration facility, uses the waste energy at low or no costs, and in effect sells the electricity to the client at a price lower than the current price paid by the client to its current electricity supplier (see Case Study 11). For financiers of the developer, the existence of the fixed assets that can serve as collateral and of regularly scheduled payments for the electricity that can be used for loan repayments are two advantages of this energy efficiency delivery mechanism.

NOTES

1. World Bank/OED (2005).
2. Case Study 3 of the Romanian Energy Efficiency Fund provides an example.
3. For details, see the Three Country Energy Efficiency Project country reports, available on http://www.3countryEE.org/reports.htm.
4. See Allen et al. (2006).
5. Also called development finance institutions (DFIs).

6. See country-specific reports at http://www.3countryee.org/publications.htm.
7. The performance guarantee could be strengthened by a performance bond, which, however, may be expensive, and manufacturers' warranties on some equipment. As an alternative to ESCO performance guarantees, energy savings insurance has been proposed; see for example Mills (2003).

CHAPTER 7

MAKING INVESTMENT DELIVERY MECHANISMS WORK

As described in preceding chapters, the challenge for increasing energy efficiency project investment is the development and sustained operation of institutional mechanisms that can deliver these investments. These mechanisms must combine arrangements for delivery of technical assessment requirements with delivery of financing. There must be incentives for the various actors in each mechanism to participate. The need for proper incentives for all actors is common sense. However, with many actors typically involved in different ways, development of constructs that provide incentives to all of the key players is often difficult and often is not addressed systematically, causing operational failures.

This chapter summarizes lessons learned from development and operation of a series of different types of institutional mechanisms for delivery of energy efficiency investments, drawing on the 13 case studies presented in Part II.

Two basic principles are first suggested below. Experience in implementation of broad categories of institutional investment delivery mechanisms—lending through local financial institutions, use of ESCOs, and DSM—is then introduced.

BASIC PRINCIPLES

Experience over the last decade in the development and operation of mechanisms to deliver energy efficiency investments, reviewed under this project or in other World Bank Group efforts,[1] suggests that at least the following two principles may best be considered in the development of new programs and major projects.

A careful diagnostic review of the local institutional environment should be conducted at the outset, followed by development of customized institutional approaches. A country- and sector-specific diagnostic review of the institutional environment in place is necessary to provide the framework for the design of institutional mechanisms that can work in that environment. For major undertakings, this should include the following reviews of the institutional environment: (i) the financial sector; (ii) local capacities for technical assessment work, including project development, and their organization; (iii) the energy efficiency market; and (iv) the role of government (and international donors) in the energy efficiency arena (including policy, regulation, and program development and implementation). In addition, knowledge of local contractual frameworks and customs is important. Although much of this knowledge may be gained only in practice, there is a growing literature on institutions, finance, and economic development upon which practitioners can profitably draw—particularly for the larger and faster-growing countries. Nevertheless, local expertise is essential for completion of diagnostic reviews, as only local experts are likely to understand the on-the-ground implications of local institutional frameworks.

Institutional mechanisms for delivery should thus combine basic principles and ideas from broader experience with customized adaptations to the local institutional environment. A variety of approaches that embody this idea are presented in this chapter and Part II, but the best choice depends on the problem and setting at hand.

This principle may appear to be common sense, but it is still frequently ignored. When a given investment delivery mechanism has worked in one setting, there is a natural tendency, both in international organizations and among in-country practitioners, to try to transplant that success to other settings. Although sharing of experience, lessons learned, and new ideas is very important, simply transplanting institutional solutions to a new institutional environment, without open-minded and systematic redesign and adjustment, is a high-risk strategy. An example is the development of training courses for developing country practitioners on how to implement institutional delivery mechanisms for energy efficiency investments, such as ESCOs or utility DSM, based on models developed and successful in industrialized countries. Where the objective is to foster new ideas, such efforts may be useful, but where the objective is "how to copy us," implementation results have been consistently poor. Case Study 7 (China ESCOs) provides an example of the difference.

End users should face commercial terms. Financing and technical assessment services to end users should be provided on commercial terms, as the financial viability of energy efficiency investment should be strong, and commercial terms provide the only foundation for sustainability in the market. As described in chapter 1, the financial returns of energy efficiency investment projects are typically high; where they are not, issues of energy pricing or the project identification and prioritization mechanisms need to be addressed. The challenge seen by the study team is to develop institutional mechanisms to deliver commercially these financially attractive, but unimplemented, investments. Reliance on end-user subsidies to sweeten the market introduces issues of the sustainability of the subsidy resources. Often, end-user subsidies underpin programs, but subsidies prove difficult to maintain over the long term as the total market size exceeds the amount of subsidy available, and programs collapse when the subsidies end. In addition, market distortions caused by subsidies can easily inhibit growth of commercially sustainable solutions in the market—making the subsidized program actually counterproductive for achievement of the goal of sustainable energy efficiency investment.

Provision of concessional financing to buy down the costs of starting up new institutional mechanisms is another matter. In the opinion of the study team, it is often a good use of public funds to support intermediaries in the development of new project delivery mechanisms. Options for this include provision of concessional finance to subsidize initially high transaction costs during development phases, to buy down risks involved in program startup, to develop and pilot new approaches and products, and to develop the wide range of capacities needed for program implementation. One example is to initially provide grants or contingent grants for energy audits that new ESCOs are frequently unable to bear up front.[2] However, these are transitional measures to help put new, commercially sustainable mechanisms in place.

The study team recognizes that opinions do vary concerning use of subsidies in promoting energy efficiency investment, as well as the definition of "transitional." At times, difficult choices are involved. In addition, there may be other, associated benefits for which subsidies may be justified. Case Study 6 (Lithuania) provides an example where investments to improve the thermal integrity of existing buildings were commercially not viable (for example, payback periods on investment averaged some 17 years), but end-user subsidies were provided, successfully increasing end-user uptake. Project supporters in that example argued that the investments have other major, nonenergy cost savings benefits, in terms of standards of living and housing cooperative development, for which the subsidies are justified. The government also moved to sustain the subsidies, using its own resources, after the development project was completed.

ENERGY EFFICIENCY LENDING THROUGH LOCAL FINANCIAL INSTITUTIONS

Commercial Banks

As described in the previous chapter, development of energy efficiency on-lending programs in local commercial banks, if possible, offers higher prospects of program sustainability and large-scale impact. Suggestions based upon the experience of the study team on how to make programs with commercial banks work are as follows:

Design of major operations might best begin with partnering with the financial intermediaries and catering to their business

approach and market development strategies. New energy efficiency project delivery mechanisms developed directly as an outgrowth of the UNF Three Country Energy Efficiency Project all have begun with partnerships with the financial institutions as the starting point. This includes the Indian SME bank lending programs, the planned new World Bank program to develop major energy efficiency lending programs in several Chinese banks, and the evolving development of a lending/guarantee facility for energy efficiency projects in Brazil, led by the BNDES. In its new, specific energy efficiency operations, the IFC also is concentrating on designing programs with local commercial banks that build on their specific market and business strategies. With this approach, project development begins with identification of local financial intermediaries interested in energy efficiency lending, followed by design of energy efficiency lending programs that conform to the specific business interests of those financial intermediaries. The financial intermediaries should select the institutional arrangements for technical assessment that best meet their needs and business preferences. Box 7.1 demonstrates, by way of a negative example, the problems that can arise when banks are not partners in program design at the beginning.

BOX 7.1 One Example of a Failed Project

Local technical consultants, combined with highly skilled international consultants, completed a major donor-funded effort to identify and pre-appraise about 15 major energy efficiency investment projects in industry in a country in economic transition during the mid-1990s. The projects identified, and the technical and financial assessment work, were of high quality. In coordination with the donor supporting the technical assistance, an MDB sought to provide a line of credit to finance the implementation of the projects.

The MDB approached the government for help in identifying a suitable bank to undertake the necessary financial intermediation for the project—accepting the line of credit for the support of the identified subprojects. The government assigned the project to one of the major state-owned banks. At project negotiations, the parties all agreed to the basic arrangements and the MDB approved the line of credit.

(continued)

> **BOX 7.1** **(continued)**
>
> As the project moved into the implementation phase, the MDB discovered that the local bank was unwilling to accept any risk associated with project implementation, arguing that it had not been involved in selection and assessment of the subprojects, and therefore could not accept responsibility for implementation results. The local bank also had no capacity to evaluate energy efficiency projects, and little interest in shouldering the major burden of developing that capacity. Finally, a simple "agent" arrangement was agreed, whereby the MDB agreed to assume the credit risk of the subprojects, and the local bank agreed to act solely in a loan processing capacity, for an agreed fee. Even after this agreement was reached, however, the local bank failed to move on any of the transactions. After many months, the line of credit was canceled without any disbursement of funds, and the subprojects identified remained unimplemented. Local bank officials, interviewed later, noted that they had little interest in lending to the types of industrial enterprises identified, and little specific interest in energy conservation investments per se. They explained that their main focus at the time was on the restructuring of their bank from a fully state-owned operation to a joint-stock, more commercially oriented operation. They had agreed to the arrangement only at the urgent request of the MDB and the government.
>
> *Source:* Authors.

Dealing with the incentives of banks to participate in energy efficiency lending. As described previously, there is nothing special about lending for energy efficiency for most banks—by itself, energy efficiency is a relatively narrow market. Many banks may be flatly uninterested. Others, however, may see energy efficiency lending as one product or tool that fits into other broader strategic objectives or marketing plans. In the many discussions between bankers and energy efficiency practitioners held under this Three Country Energy Efficiency Project, some of the following possibilities surrounding energy efficiency lending sparked the interest of the bank representatives:

- Use as one of several tools to enter or strengthen a bank's position in a specific market or business line. Examples include SME lending, lending to residential housing cooperatives or companies, targeted industrial subsector lending, development of project packages for strong industrial clients that include technical value added as well as customized financing, and development of medium-term loan tenure business (that is, 3–5 year loans).

- Development of an additional type of financial product, focusing on operating efficiency improvement, for existing good customers, as an extra service to offer, complementing and further strengthening existing lending relationships while improving the quality of their assets by improving the bottom line performance of their borrowers.

- Gaining increased knowledge of new appraisal and risk management techniques from international partners involved in promotion of the energy efficiency lending business.

- "Good citizen" benefits of energy efficiency lending. This is rarely a major point, but it may add some interest. Especially subsidiaries of major foreign banks often include broader corporate environmental or social responsibility goals into their lending policies. Lending for environmental improvements in clients' facilities can be substantial, and those banks may consider including credits for energy efficiency improvements.

External support to assist local commercial banks typically is also very important to help launch new or expanded energy efficiency lending businesses and to help such businesses to gain traction. Some options for this support include the following:

- Provision of technical assistance, especially on grant terms, to develop the institutional mechanism for a new lending program. Support for the initial operation of the arrangements for technical assessment work (including pipeline development and specialized project appraisal) can often play a critical role. Additional support for financial product adaptation and capacity building within the banks also can be critical.

- Use of partial risk guarantee programs, backed with public funds, to help defray perceived high initial business risks (discussed further below).
- Use of performance incentives. Some international financial institutions are offering incentives to banks for completion of qualifying energy efficiency lending, as part of line-of-credit operation packages, as a way to jumpstart such business. This requires sources of concessional finance, however. Examples are lines of credit for energy efficiency lending by the EBRD in Bulgaria and the IFC in Russia.

Often, just the existence of a specific, formal initiative can play a key role in fostering uptake of energy efficiency lending programs by banks, especially if the programs or initiatives are high profile or if international agencies are involved.

Integration of institutional arrangements for technical assessment work with the financial intermediation of the banks is essential. Arrangements to meet the technical assessment requirements of the bank, as well as the client, must be integrated into the operation of the energy efficiency lending program. It is strongly recommended that the institutional arrangements that the banks must rely on for due diligence be developed and controlled by the banks.

Two examples of new energy efficiency lending programs by local banks, both developed through the Three Country Energy Efficiency Project, are a series of programs in five Indian banks for energy efficiency lending to SMEs, and a proposed new World Bank line of credit operation in China.

The SME energy efficiency lending programs of Indian banks. Case Study 5 describes a series of new, small lending programs developed by the following Indian banks participating in the Three Country Energy Efficiency Project: SBI, Canara Bank, the Bank of India (Mumbai), Union Bank, and the Bank of Baroda. These banks developed their loan programs internally, but with some consultation with energy efficiency practitioners, and in some cases, a little support from IREDA in arranging and financing technical assessment work. In this program, their main challenge is in dealing with small loans for many enterprises. Generally, the new programs have been rolled out as part of existing, broader technology upgrading and modernization lending lines for SMEs. A major issue for all of the banks is how to minimize transaction costs in meeting technical assessment requirements. This is particularly difficult where initial experience has shown that there rarely are off-the-shelf technical solutions. Even if a large number of enterprises can benefit from the same or similar solutions, some of those solutions require up-front development. The institutional mechanisms being developed are described further in the case study, but a common approach taken by the banks is to cluster potential clients, either geographically or by targeted industrial subsectors, allowing concentration on just a few technical solutions rolled out among as many enterprises as possible.

The proposed World Bank energy efficiency line-of-credit project in China. Consultations between the banking community and energy efficiency practitioners supported by the Three Country Energy Efficiency Project resulted in agreement between the government and the World Bank to develop a US$200 million World Bank line-of-credit project for approval in late 2007.[3] The objective of the project is to develop sustainable and commercial energy efficiency lending programs within a series of major Chinese banks. Local banks will be responsible for all aspects of project implementation, including subproject identification, appraisal, and definition of loan terms. World Bank funds will be onlent to each bank to provide a line of credit for part of the financing. Institutional arrangements for technical assessments are under development, and will be critical. The targeted market for the program is large-scale energy conservation projects in major industrial enterprises, which will require customized, sophisticated technical solutions. A major program of technical assistance effort will be required, for which GEF and/or other international donor support is being sought, for program and financial product development as well as to support technical assessment arrangements.

Loan Guarantee Programs

Partial-risk loan guarantee programs supported by international financial institutions have shown some success in recent years in jump-starting energy efficiency financing programs through local financial institutions. This success has generated considerable interest. However, loan guarantee programs are not a broad panacea that can solve all the difficulties typically faced in efforts to expand energy efficiency investment. This instrument is designed to defray part of the risks of loan repayment for energy efficiency loans, and is particularly useful where local financial institutions attach additional risks to business concepts of energy efficiency because they are unfamiliar with the concepts or with specialized means to mitigate those risks. The instrument also may provide a useful platform for delivery of a broad package of assistance to financial intermediaries, including technical support for development of energy efficiency loan products or for development of institutional arrangements for meeting technical assessment requirements. In some cases, loan guarantee programs are not appropriate because perceived high repayment risks are not a main barrier stifling energy efficiency lending, but rather other problems are, such as lack of efficient means to meet technical assessment requirements. In these cases, support should address the other problems. In many cases, the financial sector is beset with wide problems, which a partial risk mitigation effort cannot solve. World Bank Group experience has shown that partial loan guarantee products are useful only where the banking system functions fairly well, banks can easily avail themselves of medium-term funding, and the fundamental conditions that would allow energy efficiency lending to prosper are already in place. In these cases, only a concentrated push is required to overcome perceived risks and put expanded efforts in motion.

Even with loan guarantee programs in place, end users should still satisfy loan conditions that are commercially viable—for example, where sufficient security is provided by the borrower to the banks. Loan structures and repayment security measures in the presence of partial guarantee programs may be designed to better meet the needs of energy efficiency projects, but clients should not expect participating banks and their guarantors to accept unsecured arrangements that are not commercially viable or that are backed with insufficient credit.[4]

Two of the best-known energy efficiency loan guarantee programs are the GEF/IFC Hungary Energy Efficiency Co-Financing Program (HEECP) (Case Study 2), and the China ESCO Partial Loan Guarantee Program (Case Study 1) in the GEF/World Bank China Second Energy Conservation Project. The Hungarian program was launched in 1997 and is now in its third phase. The program has been quite successful in catalyzing energy efficiency lending businesses in Hungarian commercial banks and has been expanded to other countries in Central and Eastern Europe. The Chinese program was launched in November 2003 and has already backed US$57 million in new ESCO energy performance investments during 3.5 years of operation. It has achieved goals set so far, but achieving greater leverage and developing new products to widen the influence of the program in the market would be important to support the growth of China's ESCO industry during the second half of the seven-year program.

While the two programs share some common points, they are quite different, due in part to different objectives, but also to different designs that reflect in part differences in the institutional environment of the financial sector in the two countries:

Program objectives. The Hungarian guarantee program has sought to build a sustainable commercial lending business in Hungary for energy efficiency investments across a range of sectors. The Chinese guarantee program is narrower, aiming to expand domestic energy efficiency investment specifically through energy performance contracting, only through ESCOs, leaving efforts to expand direct energy efficiency lending from banks to clients to other programs.

Program structure. In its Hungarian program, IFC directly enters into overall Guarantee Facility Agreements with participating banks and issues partial loan guarantees for each transaction (or a portfolio of transactions), backed by GEF and IFC resources. Technical assistance is provided both to the banks and to project developers. Much of the success of the program has been built through the relationship between

IFC and the banks on all aspects of development of energy efficiency lending businesses. Implementation of the Chinese program, in contrast, has been entrusted to China's largest guarantee company, I&G, backed with GEF resources provided through the World Bank and the Government of China. I&G serves as a focal point for ESCOs seeking loan financing and helps to structure transactions. It enlists local banks and helps them understand the ESCO business, and it provides credit guarantees to the banks based on I&G's due diligence reviews of the ESCOs and their projects.

Local banking environment. In Hungary, competition in the banking sector is strong, interest rates are low relative to many developing countries, and the banks are quite sophisticated. In China, competition in the banking sector is growing, interest rates are relatively low, but sector reforms are still under implementation. In particular, until recently, interest rates have been restricted to a narrow band by the government, creating little or no price incentives for banks to incur risk.

Guarantee coverage. In the Hungarian program, IFC's guarantee coverage has been as low as 35 percent of loan amounts. In China, where guarantee companies have traditionally guaranteed 100 percent of loan amounts (incurring risks or developing new project lines, with which banks have less experience, for an additional fee from clients), I&G has been guaranteeing 90 percent of loan amounts. As bank sector reform in China further proceeds, however, I&G's coverage is expected to be reduced.

Technical assessment arrangements. With technical assistance support under the program, IFC and its bank partners have utilized ESCOs, leasing companies, and equipment and energy efficiency service vendors for a substantial portion of deal origination and for subproject aggregation. In China, I&G works in strategic partnership with China's EMC (ESCO) Association for identification of a substantial portion of new projects and project concepts. I&G also engages energy efficiency experts to work together with its own appraisal department on the technical aspects of project review.

Labor intensity. Both programs have required an enormous input of staff time from all of the institutions involved, as well as major, multiyear technical assistance efforts, to achieve traction in the marketplace. GEF support, especially during initial phases, has made such human-resource intensive investment possible.

Use of Development Finance Institutions and Special Revolving Funds

Another common approach to expand loan financing for commercially viable energy efficiency investments is to use existing or new specialized institutions or funds, developed specifically for such purposes. A number of countries have created special DFIs for financing energy efficiency and alternative energy. Examples include IREDA and CECIC, which are both publicly owned. Rather than adopt the formal DFI approach, other countries have developed special energy efficiency loan funds, as special legal entities governed by boards or foundations representing both the public and private sector. The Romanian Energy Efficiency Fund (FREE[5]) and the Bulgarian Energy Efficiency Fund (BgEEF) are two examples. Finally, many projects and programs include special revolving funds, which are typically fairly small, for directing credit for specialized purposes, such as defined types of demonstration projects.

Two case studies in Part II were developed to provide examples of these approaches. Case Study 4 outlines the energy efficiency lending business of IREDA. IREDA is one of the largest DFIs specializing in renewable energy and energy efficiency in the world, with a cumulative total of over US$1.6 billion in loan commitments. IREDA added an energy efficiency lending business to its renewable energy business in 2001. Case Study 3 describes the institutional mechanism of the Romanian Energy Efficiency Fund, which began operation in 2002.

Advantages of these specialized entities or funds approaches include an ability to concentrate attention to the specialized task of energy efficiency lending and an ability to offer clients a "one-stop shop." Typically, these programs combine financial intermediation with project development and other technical assessment functions in a largely hierarchical approach. Financial products can be designed specifically for the energy efficiency business, and entities can provide value added on technical aspects as well as financing.

Where the local financial sector is under stress, or in the midst of major transitional reforms and restructuring, use of such specialized institutions outside of the banking sector may at times present the only option forward. This was the case when FREE was designed in Romania. However, creation of such funds outside the banking system can deter and impede development of commercial, sustainable, market-based financing instruments. Thus, creation of revolving loan funds (and their impact in dynamic markets) when executed outside the commercial banking industry should be very carefully considered.

Although it may present a perceived advantage at some point, separation of these entities and funds from the commercial banking sector also carries disadvantages and major risks. In the case of IREDA and FREE, this was understood at the beginning, and considered in program design. Yet difficulties remain, as markets tend to change quickly and relationships between such entities and commercial banks are complex.

Often DFIs are created, as in the case of IREDA, with a mandate to pioneer new financial interventions and products in the national interest, which commercial banks can then replicate on a larger scale. Once successful in piloting new businesses that previously were perceived by others as too risky or untried, the DFI may not be in a position to compete with commercial entities, and actually may be expected under its mandate to withdraw from such competition. Where entities such as IREDA are expected to be profitable, the implicit requirement to shoulder the high commercial risks of new development, without being able to take long-term advantage of the lower risk and lower transaction costs through scale-up of successful developments, presents a difficult dilemma.

Specialized revolving funds targeted for very particular needs in broader program packages may at times be beneficial; but care must be taken to define the objectives clearly and to think through the operating mechanisms from the start. Sometimes revolving funds are added into development projects as part of a broad idea that some type of credit assistance to clients is necessary in the program; however, the fund details have not been defined. Many such efforts fail. Program planners should ask a series of important questions up front: Is the objective only to fund a series of demonstration projects for eventual replication, or is it to develop a sustainable financing mechanism? Who is responsible for shouldering credit risk? What are the incentives for all of those involved? Is it truly necessary to provide loan financing under the project, or can loan financing from others be mobilized (by providing special incentives), and the complexities of special loan financing be avoided? It should also be recognized that developing a functional understanding of the detailed legal, fiduciary, and commercial aspects is relatively time-consuming but critical for developing a successful program.

ENERGY SERVICE COMPANIES (ESCOs)

ESCOs can be an important institutional mechanism involved in delivery of energy efficiency investment. The term ESCO (or EMC or EMCo—Energy Management Company—in China) may mean different things to different people.[6] At times, the term is used simply to describe companies involved in completing technical assessment work, especially energy audits. In this book, ESCOs are defined to include any company using energy performance contracting as part of energy efficiency investment transactions. An energy performance contract (EPC) in the ESCO business may be broadly defined as a contract between the ESCO and its client, involving an energy efficiency investment in the client's facilities, the performance of which is somehow guaranteed by the ESCO, with financial consequences for the ESCO. Technical assessment work is a key part of the work of all ESCOs. Although there are many variations in the business models used, a key distinction between ESCO business models involves whether or not the ESCO provides any financing for the investment projects it develops. An example of a business model where the ESCO provides financing is the "full service" ESCO model employed in China (Case Study 7), where the ESCO identifies, designs, finances, and oversees installation and commissioning of projects in the facilities of the client, and receives compensation in the form of a share of the energy savings achieved over a defined period, according to a "shared savings" EPC. An example of a different business model practiced in the United States is provided in Case Study 8, where the ESCO only assists in arranging financing, but identifies, designs, oversees installation and commissioning of projects, and guarantees the energy savings to the client over a defined period, according to a "guaranteed savings" EPC (see also figures 7.1 and 7.2).

ESCOs that provide financing to clients may be viewed as a partial energy efficiency investment financing mechanism, operating at the retail level. These ESCOs serve as project aggregators, to which financial institutions may provide financing for a package of projects, and reduce their direct involvement with end users. ESCOs that do not provide financing to clients may best be categorized as technical assessment and engineering services entities, although their guarantee of the energy savings from investments may be a critical element for closing the related financing contract between a financier and the client.

ESCOs exist in each of three large countries included in the Three Country Energy Efficiency Project. The ESCO industry is largest in China, where over 60 ESCOs were operating at the end of 2006, with

Figure 7.1. Shared Savings EPC Model

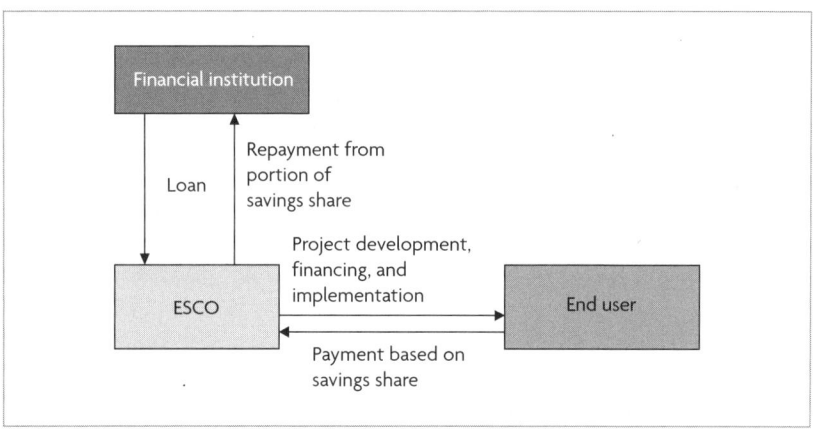

Source: Authors.

Figure 7.2. Guaranteed Savings EPC Model

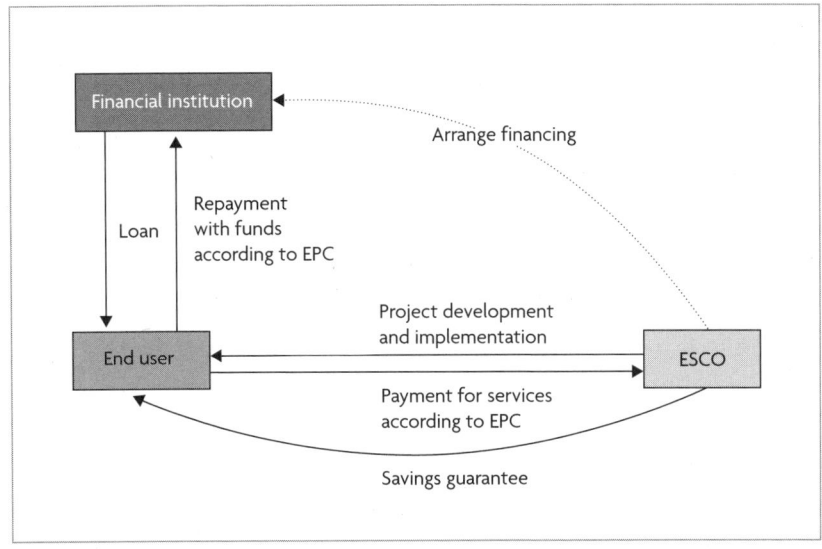

Source: Authors.

total annual energy efficiency investments through energy performance contracting of about US$280 million and growing. In Brazil, there are dozens of engineering firms providing some energy efficiency services, of which fewer than 25 have sufficient project flow to permit more diversified professional capacity and can potentially enter into EPCs. Investments of these companies are estimated to be on the order of US$40 million per year. In India, various ESCO models are well known among energy efficiency practitioners, and about 20 ESCOs are operating with some type of energy performance contracting business model. However, most ESCOs are very small, and the ESCO industry has yet to gain a substantial footing in the Indian market. National ESCO industry associations are active in all three countries, as key industry advocates (ABESCO in Brazil, EMCA in China and the recently formed ICPEEB in India).

The ESCO model offers an appealing, market-based approach to energy efficiency project development. Especially where ESCOs finance projects, they offer an attractive proposition to clients: the ESCO will do the project design and management, shoulder the specific project technical risks, provide off-balance-sheet financing, and seek payment only from the cash flow of energy savings generated. Even if financing is not included, the ESCO can play a key role by assuring both the client and the financier that the cash flow from the project will materialize. ESCO business models cater to private sector companies. In China, the models adopted have proven quite profitable to date, and potential profitability is attracting many new entrants.

Despite its appeal, the ESCO business model also has proven very difficult in many developing country environments to launch and develop in practice. Market-based contractual systems are immature in many emerging market countries. Energy performance contracting, on the other hand, is a sophisticated instrument requiring relatively complex contractual arrangements with which financiers, clients, regulators, auditors, and tax collectors are typically unfamiliar. It is also very difficult for the small, technically oriented firms that often provide the seedlings for new ESCO businesses to gain access to credit or to be taken seriously by financiers when trying to arrange credit for others. With these problems in mind, the study team summarizes the lessons it has learned and provides some suggestions to others, drawing especially from the cases of Brazil, China, and India:

- **The ESCO model is not a magic bullet to solve problems in delivering energy efficiency investment.** In view of the growing success of ESCOs in China, it is clear that ESCOs can play an important role, if local institutional environments are suitable. However, given the complexities of ESCO industry start up, it may be appropriate in some countries for priorities to be placed elsewhere. In any event, ESCOs represent only a partial solution at best. These businesses usually operate only in some corners of the market—for example, in building market segments or with certain types of relatively simple industrial projects; projects with technical assessment requirements fitting the ESCO's capacity; and, usually, relatively small projects. Moreover, the ESCO model does not solve basic problems of delivering energy efficiency project financing. Even when ESCOs provide financing to clients, the difficulty of the ESCOs themselves in obtaining project finance is an area of concern.

- **ESCO project financing issues should be considered up front in any serious effort to promote local ESCO businesses.** Programs to provide only technical assistance to prospective companies on possible business approaches, without attention to solving practical financing issues, have not proven very helpful. It is difficult for a fledgling ESCO industry to develop without mechanisms to obtain financing for projects. Again, financing remains central to all variations of the ESCO model of energy efficiency delivery.

  In China, the ESCO industry is developing through three phases. The first phase began when the ESCO concept was completely unknown locally. Support through financing provided by GEF and the World Bank was provided to three demonstration companies for trying out the energy performance contracting mechanism in the market. With success achieved from operating the model in the market, the second phase provided a mechanism to help new ESCOs obtain financing using the ESCO loan guarantee program. The emerging hypothesis is that, as the ESCO industry grows in strength, as banking sector reform continues, and as banks become more familiar with the energy efficiency business through the guarantee program, ESCOs in China will increasingly be able to obtain financing directly from local banks in a third development phase.

  In Brazil, ESCOs have been able to take advantage of financial resources from electric power utilities, developing projects for utility customers as one means to meet energy efficiency investment requirements mandated by ANEEL, the national electric power regulatory agency, under Brazil's public benefit wire-charge mechanism (see Case Study 9). In an environment where it is difficult for small companies to obtain medium-term credit from the banking system, cooperation with the utilities and their customers through the wire-charge program has enabled many of Brazil's ESCOs—especially the larger ones—to develop and grow their businesses. However, it has not led to developing the capabilities to obtain commercial bank financing.

  In India, by comparison, the ESCO industry does not have sustained specialized mechanisms to turn to for project finance, and the industry is having difficulty gaining traction despite the successful implementation of several pilot projects in the past.

- **Active government support for ESCO development is critical, especially during early stages.** Experience suggests that strategically placed and active government support is critical to enable a local ESCO industry to develop and gain strength. This was certainly true in the United States and Canada, where the concept first took hold. As described in Case Study 8, the ESCO industries in these two countries grew largely through support from federal and local government programs promoting energy efficiency renovation in public facilities, thereby creating a core market for ESCOs where projects were relatively easy to finance since the public sector clients were very creditworthy. Government-mandated utility demand-side management programs also added markets, in some cases specifically supporting the ESCO mechanism. In China, ESCOs were introduced and supported as a strategic initiative of the central government as part of an effort to develop market-oriented energy efficiency promotion mechanisms as the country shifted from a planned to market economy. In Brazil, the electricity utility wire-charge program mandated by the government has helped to support ESCOs (although the Brazilian ESCO industry also has expressed urgent needs for direct and targeted support). Given the difficulties experienced in developing ESCOs in countries without strong and sustained government support, the conclusion is as follows: if development of an ESCO industry in a given country is truly considered important, the government should make a concerted effort to incubate and foster development in the early phases.

    Strategic government support may focus on market creation or support for project financing mechanisms. In addition, governments can play an important role in legitimizing this nontraditional business model. The central government played this role in China, which was essential for financial institutions, local regulators, and tax authorities to accept the new business approach.

- **Roles of shared savings and guaranteed savings energy performance contracts.** Choice of ESCO business models should best be determined by the local markets, by financing practices and systems, and by institutional environments. In the opinion of the study team, there is no right or wrong model independent of knowledge of the local institutional environment. What matters is increasing energy efficiency investments. The perceived value of energy performance contracting is likely to be different for different types of customers. Some building or shopping center owners may have little knowledge of energy savings technologies and their operation, and a guarantee of the energy savings resulting from an investment may be especially important to them (figure 7.2). Some industrial clients, on the other hand, may be acutely aware of the energy savings likely from certain investments, and their main interest may be in off-balance-sheet financing, resulting in a shared savings contract with the ESCO providing the financing (figure 7.1). In the United States, for example, the high creditworthiness of institutional customers eased their ability to obtain financing directly for energy efficiency projects and created opportunities for ESCOs to move to guaranteed energy savings contracts without providing financing (figure 7.2). The lesson is that where creditworthiness of customers or their interest in new borrowings for energy efficiency are key issues, guaranteed energy savings models are likely to be difficult to use.

- **The skills needs of ESCOs.** Well-established ESCOs require a mix of skills, presenting a challenge for any company. Technical skills are mandatory, if the company is to be able to minimize technical risks and provide technical value added to customers. Business skills—especially how to market the business, sell its product, and close transactions—also are essential. Finally, financial management skills are critical, especially if the ESCO is providing financing to customers. In this case, competencies in assessing repayment risks, and in mitigating those risks both for individual projects and through portfolio management, are likely to make or break the company.

- **Capital needs of ESCOs.** Executing ESCO business models requires a substantial level of equity. This is particularly true for ESCO models where the ESCO is the borrower of funds to finance projects. However, this is also true in the "U.S. model" where substantial capital must be available to back up performance guarantees in order to make them credible in the market. Finally, the ESCO project development cycle may be quite long, requiring substantial working capital to support the ESCO for marketing, closing project financing, constructing and commissioning capital projects, demonstrating savings, and finally generating revenues through client payments.[7]

ENERGY UTILITY DEMAND-SIDE MANAGEMENT (DSM)

Although utility DSM programs were not one of the topics explicitly covered under the UNF Three Country Energy Efficiency Project, these programs do represent another option for promoting energy efficiency investments.[8] A strong advantage of utility DSM is the use of what are usually well-developed and financially strong utility institutions that have direct relations with energy users for program delivery. In some cases, as in Sri Lanka (Case Study 10), utilities may even be able to use electricity bills as the contractual mechanism to ensure repayment from customers for energy efficiency investments. A strong disadvantage of this mechanism is that most utilities do not have natural incentives to promote true energy conservation (as distinguished from load management), which results in a loss of sales of their core product. Although temporary energy shortages may spark utility interest in promoting energy efficiency for a time, especially when subject to high-profile publicity and political pressure, the fundamental economics of the utility's position as an energy supplier still tends to undermine utility interest in DSM/energy efficiency programs over time. In countries such as the United States, therefore, DSM programs were only successful on a large scale when governments and regulators made the programs mandatory and they allowed utilities to receive special compensation from electricity consumers for lost revenues. To use the DSM mechanism effectively, the utility incentive issue must be properly addressed. In recent years progress has been made in this respect, for example in the United States by decoupling utility revenues from sales volume. In addition, many states are providing positive incentives to utilities for meeting efficiency goals.[9]

Creation of mandatory regulations to pursue DSM programs is one approach, but this obviously requires strong political support and a strong regulatory system. As vertically integrated electricity companies become unbundled, following worldwide trends in power sector reform, regulatory promotion of DSM also tends to become more complex.

At times, however, it may be possible to align a DSM program with other interests of the utility to improve utility incentives. One example is to reconcile electricity DSM programs with load management interests of utilities. Another example is to develop energy efficiency service programs for customers that allow the utility to capture new customers (or retain existing ones), resulting in increased utility sales, even if energy is used more efficiently.

Utility DSM programs may align energy efficiency interests with the interests of the utility to reduce peak load requirements—the most expensive part of the load curve to service. Case Study 10, describing a DSM program of the Ceylon Electricity Board (CEB) in Sri Lanka, is fairly typical of this situation in its promotion of more energy efficient lighting technology that both reduces the sharp peak caused by the residential evening lighting load and saves kilowatt hours. Occasionally, utilities also may be interested in energy efficiency measures among certain customer categories where the utilities are losing money on electricity sales under the prevailing pricing structure—for example, through cross-subsidization by other consumer categories. DSM can provide benefits (or rather reduced losses) to utilities when customers are targeted whose tariff is lower than the cost of service. This happens frequently in the case of public lighting. Case Study 9 describes the substantial investments to improve the efficiency of public lighting in Brazilian municipalities, which has little relation to their rather small impact on the load curve. It should be remembered, however, that changes in the pricing structure may cause incentives to participate by the utilities to disappear virtually overnight.

Case Study 12 presents an example in Brazil where the subsidiary of a natural gas supply utility provides both financing and technical support to customers to adopt more energy efficient solutions as part of a shift to natural gas use, including investment in cogeneration facilities for both cooling and own-use power generation under a chauffage-type arrangement. The business results in increased natural gas sales for the parent company, by adding new customers, as well as profits from the chauffage-type investment and operation business.

NOTES

1. For example, World Bank 2004.
2. For more details see World Bank 2004.
3. A case study has not been prepared as the proposed institutional mechanism is not yet operational.
4. This was an issue when the China ESCO Loan Guarantee Program was launched. Some clients were initially aghast that evidence of strong creditworthiness, and/or counter-guarantees or other security measures, were still required under the program (even if approaches were more "ESCO-friendly" than standard bank approaches).
5. The acronym comes from the Romanian name, "Fundul Roman pentru Eficienta Energiei."
6. See for example the list of different ESCO business models in World Bank 2004.
7. See chapter 6 about ESCOs' problems to attract equity investments.
8. See, for example, ESMAP 2000.
9. See, for example, Harrington, Murray, and Baldwin 2007.

CHAPTER 8

CONCLUSIONS AND RECOMMENDATIONS

From the previous pages, it is obvious that expanding energy efficiency investments in developing countries is not simple or easy. The past record of efforts to increase energy efficiency investments among end users is mixed, and there are ample cases of initiatives that could not get off the ground or failed to meet expectations in implementation. The key advantage for energy efficiency investment development is the economic and financial fundamentals of the projects themselves—there are many projects with strong financial returns. The problem is that these cost-saving investments are not being made through the market in numbers in any way approaching their potential, especially in fast-growing developing country economies. To foster a more substantial take-up of these publicly desirable investments, therefore, efforts must focus on development of improved energy efficiency investment delivery mechanisms that can operate in the market and help accelerate investment to levels more closely approximating the potential. However, development of improved investment delivery mechanisms that will work well over prolonged periods is a difficult proposition, as borne out by the difficulties experienced in energy efficiency operational work.

Development and operation of energy efficiency investment delivery mechanisms is an institutional development issue, and energy efficiency financing programs and projects must recognize this clearly. This requires a major effort to ensure that the planned institutional solutions match the institutional environment in which they are expected to function. In addition, new institutional mechanisms cannot be expected to develop and grow overnight, and therefore sustained efforts are required. The successes achieved in developing energy efficiency lending schemes in Hungary or ESCOs in China, for example, have been the result of about eight years of persistent effort.

In most cases, steady and strategic government support is a very important enabling factor for the type of institutional development required to truly improve delivery of energy efficiency financing. In addition, where international financial institutions have played a successful role assisting in energy efficiency financing development, this also has been through multiyear, concentrated programs and efforts. Experience has shown that there simply is no magically fast or simple approach.

SUGGESTIONS FOR EACH OF THE THREE COUNTRIES

One clear message from the experience of the UNF Three Country Energy Efficiency Project is the importance of establishing and maintaining practical, operationally focused dialogue between the banking community and energy efficiency practitioners. Launching this dialogue was probably the most important contribution of this project to the energy efficiency programs in each of the three countries. The core groups in each country are now particularly interested in building further upon the platform created, especially in specific ongoing or upcoming projects, and the specific recommendations of each group for follow-up work reflect this.

Another clear conclusion from the Three Country Energy Efficiency Project is the central importance of strategic government support to more aggressively promote new energy efficiency financing mechanisms. Strategic government support does not necessarily involve the commitment of large amounts of public funds. Rather, what is critical is for the central governments of Brazil, China, and India to utilize their convening power and certain strategically focused but sustained institutional development support interventions to enable new concepts to gain operational footholds and then scale up.

Suggestions for each country are summarized below, and further detailed in the individual country reports.

Brazil

As of the end of 2006, Brazil had experienced the greatest difficulties among the three countries in developing energy efficiency financing mechanisms involving the banking sector. Although interest rates have come down substantially, they are still quite high and the current bank lending environment is not conducive to SME loan financing. At the outset of this project, the gap in understanding between energy efficiency promoters and financing agents about the other's requirements and opportunities also was particularly noticeable. Over time, this gap has begun to narrow and new ideas have been generated. It remains critical to build further on this platform and to move into implementation of investment programs. The study team suggests two central means of doing this. First, BNDES should proceed with implementation of a pilot energy efficiency financing program, as means to engage the banking sector at a practical, operational level with concrete investment projects, however small. Second, the government should consider undertaking, possibly together with support from the World Bank and other donors, a strategic review of energy efficiency investment development, as a means to consider the best way to respond to ever-mounting energy needs over the next 5–7 years.

There has been ample discussion of various proposals for energy efficiency financing mechanisms, especially guarantee funds, over the last several years. What is important now is to implement a series of pilot investment projects, even small ones, which is the best way to address and overcome specific operational issues and gain confidence in an operational framework. The proposed pilot risk-sharing scheme between the BNDES and several banks, called PROESCO, is one example. Implementation of this scheme, or a similar pilot scheme, could substantially move the energy efficiency financing agenda forward. What is needed are pragmatic efforts to test mechanisms with real projects, and then adjust where necessary, rather than attempt to design the optimal mechanisms and resolve all potential operational issues up front. It is recommended to launch an effort to support several small investments in relevant financial mechanisms, engaging the institutions involved and developing solutions to operational issues based on real projects. Besides credit guarantee mechanisms, other possible mechanisms for pilots to enhance access to financing are performance bonds and a fund to buy energy efficiency project receivables.

These pilots can be implemented within a fairly short time. For example, PROESCO could begin seeking projects within several months of being approved by the BNDES. Other possible pilots would take longer, but it is important for the government and especially international donors to begin to prepare means to support the successful implementation of these pilots. Project developers, consumers, and financial agents would be going up a steep learning curve.

Furthermore, the study team believes that now is an important time to undertake a substantial review of the status of energy efficiency investment in Brazil, and of strategic means to spur national-interest investment over the medium-term, particularly from the private sector. So far there is no policy framework in Brazil to support this kind of initiative. Energy efficiency investments are an important aspect of improving productivity and competitiveness, reducing operational expenses of public sector facilities, and helping address long-term strategic energy supply security issues.[1] In addition, as promotion of energy efficiency has become one of the highest priorities in the renewed international focus on clean energy development and mitigation of global climate change, a strategic review with follow-on support by the government could strengthen Brazil's position as a leader in this field. Brazilian enterprises could benefit more substantially in global carbon emission reduction efforts if responses are well organized and strategically focused.

In undertaking such a review and considering strategies for the future, it is important to view energy consumption and efficiency in a holistic manner, even though electricity and fuel efficiency issues have been managed separately in Brazil for historic institutional reasons. The distinctions are artificial for the energy efficiency business, as energy efficiency investments focus on the issue of reducing energy costs, regardless of the energy form. In addition, the efficiency of water use could also be included.

The following key issues are recommended for focus: (i) current demands and trends for mobilizing energy efficiency investment in the private sector, especially in the industrial and commercial sectors, and tools to facilitate further investment using medium-term debt financing; (ii) future use of the electricity wire-charge to promote energy efficiency investment, and how to improve its operation; (iii) identification of pricing issues hampering energy efficiency investment, and recommendations to overcome them (one important area for review may be the current time-of-day tariff for electricity); and (iv) means to accelerate energy efficiency investment in the public sector, including, in particular, public buildings and water utilities. Further discussion of these topics is presented in the Brazil country report.

Finally, it is important to continue to support and follow up some of the specific initiatives of the last few years. One example is support for Brazil's ESCO industry, which has survived the test of time and a wide range of operational difficulties to remain an important, market-oriented player in the energy efficiency arena. Another initiative to pursue is finally breaking through the procurement hurdles for energy efficiency projects in the public sector.

China

During 2005 and 2006, the importance of achieving results in energy efficiency has achieved more prominence on the national stage in China than perhaps ever before. The challenge in China is to deliver the actual results through policy and program implementation.

China's government has set an ambitious, official target to reduce the energy intensity of the country's GDP by 20 percent during 2006–10. This translates into a target to achieve energy savings of at least some 600 million tons of coal equivalent by 2010. Although some of these savings may be achieved through changes in the structure of the economy, perhaps half will need to come from technical improvements in the efficiency of energy use. The government has set targets for various sectors and regions in a strong, organized effort to ensure the technology uptake and mobilization of investment for energy efficiency. Government efforts include a series of policy initiatives to foster technology development and dissemination, assess progress in key sectors and enterprises against benchmarks, and provide various fiscal incentives. However, the actual investments and hence attainment of results must come mainly through the market. Unlike the 1980s or early 1990s, the government now has only an indirect influence by providing a solid enabling environment for market-based actors.

In this context, efforts to promote increased energy efficiency lending in the banking sector have taken on increased urgency. Yet, energy efficiency lending to date has been much less than once hoped. Still undergoing reform, many Chinese banks have tended to be less nimble in responding to new lending opportunities than some counterparts in other countries, and the energy efficiency business is largely new to them. New initiatives to develop energy efficiency lending businesses in selected banks are therefore very important. These include the planned new and large World Bank line of credit to several banks that is aimed specifically at the development of new commercial energy efficiency lending programs, a new GEF-supported initiative by IFC that includes a partial-risk guarantee program for energy efficiency lending by participating banks, and several smaller initiatives by bilateral assistance organizations. The challenge is in the details of implementation of these new and difficult initiatives, which will require prolonged and intensive effort by all parties.

China's rapidly growing ESCO industry is poised to make a significant contribution to the national energy efficiency agenda. Perhaps the two biggest challenges at this point are to further the effort to create sustainable systems for ESCO financing from the banking sector and to ensure stability in the ESCO industry as it grows quickly. Whereas Chinese ESCOs have proven agile in finding sources of financing outside of the banking sector, and the ESCO Loan Guarantee

Program has begun to lay a foundation for bringing the banks into the business, direct domestic bank lending for ESCO projects is just beginning. Use of the Guarantee Program and other means to foster increasingly strong and direct business relationships between the ESCOs and the banks is one of the most important areas for work over the next several years. Another area for focus is for ESCOs to foster healthy operations and a good reputation as the industry proceeds through a period of very rapid growth. Some type of ESCO accreditation program may be a useful tool to achieve this goal.

Although on the national agenda now, one area where energy efficiency efforts have lagged in China is in public facilities. The potential for energy efficiency gains in existing and new public buildings remains largely untapped, such as in government office buildings, hospitals, and schools. To develop this potential will require special mechanisms adapted to government procurement requirements, government budgeting procedures, and the incentive/disincentive structures that these create for public agencies.

India

A wide range of energy efficiency initiatives are underway in India, many of them quite creative and with great promise. To generate the best results in terms of actual energy efficiency gains, the study team believes there is great need for a strong strategic review at the national level, involving the central government. This review should assess priorities for work on energy efficiency development in the coming years, and focus sustained, multiyear attention on the implementation of the policy initiatives and market-oriented investment mechanisms that can provide the biggest contributions. Such an integrated and strategic review would also be useful for establishing national priorities for support under current and new international clean energy and climate change initiatives. A review might begin at the macro level, assessing energy intensities and potential savings in different sectors,[2] and the practical, implementable investment areas that could yield the biggest benefits. But the most important point would be to establish priorities for specific programs to generate the necessary investment.

India's Energy Conservation Act provides a solid, comprehensive regulatory framework that blends both voluntary and mandatory programs that aim to increase energy efficiency activities in certain energy-intensive sectors and in the economy as a whole. However, implementation of many provisions of the Act has lagged, and there is a need to refocus country efforts on how to translate the ambitious provisions of the Act into actual results on the ground.

The Three Country Energy Efficiency Project touched on only some of the many market-oriented energy efficiency activities in India. In each, there is promise, but a strategic decision is needed whether or not to provide the sustained government support that typically is required for long term success. Future levels of government support to IREDA are uncertain, which raises questions about their future participation in the energy efficiency investment business and what overall role IREDA might best play in spurring domestic investment in efficiency.

Another question concerns the future of ESCOs in India. ESCOs as well as energy auditors have made little progress so far in reaching the wider Indian market. This is partially due to their small size and limited reach, and lack of credibility and relationships with other important actors. Support for these two groups, whose prime business deals with energy efficiency and who could become important promoters of energy efficiency investments in India, could contribute to increasing commercially based energy efficiency investments. Although ESCO development may or may not be considered a priority, it is clear from the past experience in India and internationally that India's ESCO industry is unlikely to develop significantly without sustained government support, through ESCO market creation initiatives (perhaps through new energy efficiency initiatives for public buildings) or other means.

Also to be addressed is the future of the promising new energy efficiency lending businesses being developed by the Indian banks. The launching of dedicated energy efficiency finance schemes by five major banks in India was one of the most successful accomplishments of the Three Country Energy Efficiency Project. The India core team would like to continue to work to expand the program, but this will require some level of public support, even if modest, most likely through the donor community. Proposals are under preparation. Several specific areas where follow-up efforts could yield significant returns include the following:

- Further assistance to the participating banks to refine and standardize loan applications/appraisal procedures, and minimize transaction costs. This requires specialized technical assistance to each individual bank and follow-up training assistance. Additional work on options to "ring fence" the negative cost stream achieved in energy efficiency projects for partial use as loan security also would be beneficial.

- Intensive support to the local bank branches in marketing and refining their energy efficiency lending schemes. Much work remains to be done in further disseminating information about these schemes to energy efficiency project developers and client enterprises. Further efforts also are required to improve capacities and the efficiency of arrangements for the technical assessment aspects of the lending schemes.

- Government characterization of energy efficiency projects, and not just small-scale industry lending, as "priority sector" projects would facilitate uptake in public sector Indian banks.

Finally, it is important to emphasize that while government support for energy efficiency is required, it should not result in the government stifling the activities of other actors by telling them what type of energy efficiency schemes to formulate or end-user segments to target. Actors on the ground need to make their own decisions based on their business objectives.

SUGGESTED ROLES FOR INTERNATIONAL FINANCIAL INSTITUTIONS

Many of the case studies described in Part II involve investment and technical support from international financial institutions (IFIs) such as the World Bank, the IFC, the regional multilateral development banks, or bilateral development aid agencies. The IFIs (including institutions providing bilateral as well as multilateral investment assistance) are often a source of outside ideas and experience, as well as a source of financing. Often, in large countries, the most important core role of the IFIs is to provide a formal project framework for trying something new.

Projects to develop local financing mechanisms for energy efficiency investments involving IFIs need to give full play to the advantages that these institutions can bring to any development project, in particular the following:

(i) Ability to combine investment financing and technical assistance support in multiyear packages. Investment financing support of different types enables concentration on actual implementation of the investment mechanisms, rather than just planning and training in how implementation might be undertaken. But strong technical assistance programs also must accompany investment financing support, to foster the implementation of new ideas and ways of doing things inherent in institutional change.

(ii) Ability to maintain a sustained presence. Creating and operating new institutional mechanisms requires a steady, long-term operational relationship, to follow through from design to development to start-up and finally to operational rollout. Practical problems arise all along the way, requiring sustained support.

(iii) Ability to provide international information in a usefully structured form. IFIs have access to operational information from many countries, which can be brought to bear on specific operational issues.

Use of Different Financing Tools and the Role of IFI Lending

The objective of most energy efficiency financing projects today is the development of sustainable institutional mechanisms to deliver energy efficiency investments. The problem faced in most countries is not a shortage of capital for energy efficiency investment, but lack of means to deliver existing capital to end users in suitable financial and technical packages. As shortage of domestic capital is not the problem, it follows that provision of flows of external capital also is not the solution. What is required is development of delivery systems. To achieve this, a variety of tools can be applied by IFIs. Incorporating investment support with technical assistance typically is important—not for the capital flow as such, but to focus the effort on practical, operational implementation. Lending volumes need not be big and other instruments such as guarantees or well-targeted grant support may be as useful as lines of credit.

Two types of financial support from the World Bank Group may be useful, depending upon the specific needs: either IDA or IBRD lending or guarantee support through the World Bank; or lending, guarantee, or equity investment financing support through IFC. Generally speaking, support through IBRD and IDA instruments best meets project needs if it is important for governments to be involved, whereas support through IFC will provide more nimble solutions where governments need not be involved and the focus is on commercially oriented transactions with nongovernment entities.

The extent of IFI emphasis and contribution for specific development tasks is often measured in terms of the total amount of capital lent. This is clearly a misplaced measuring tool to gauge emphasis and success in the case of IFI involvement in end-use energy efficiency investments. The success of IFI involvement in energy efficiency should be measured, where possible, in terms of the energy savings achieved from IFI-sponsored programs and activities.

The Important Role of the GEF

GEF project support for energy efficiency financing projects has been critical and beneficial over the last decade. As shown in the case studies, many of the major energy efficiency financing projects in the developing world with multilateral international assistance have involved GEF financing support. The GEF's OP 5 provides a good framework for development of new mechanisms to deliver commercially viable energy efficiency investments. A particularly important aspect of GEF's support is that it includes funding for investment, which can be combined with technical assistance. GEF grant financing has underpinned the introduction and development of new mechanisms and pilot projects, in an effort to cover part of the initially high transaction costs of schemes and especially to help cover initial risks. Over the years, the skill of operational staff in using GEF grant resources as a catalytic and leveraging tool has increased, such that the energy efficiency gains per unit of GEF investment in a number of new energy efficiency financing projects are strikingly high.

OPERATIONAL SUGGESTIONS ON DEVELOPMENT OF NEW PROJECTS

This book provides an analytical framework for addressing issues surrounding the financing of high-return energy efficiency investments. It also outlines the ingredients for successful projects in this area, and summarizes the approaches that have been used. Lessons learned from a series of examples are included. The authors hope that the analytical framework provided and details concerning project implementation experience will be a useful contribution to those considering development of specific new projects. In closing, the study team offers the following broad suggestions to developers of new projects and programs.

Our review of recent experience suggests that the three biggest causes of operational failures in energy efficiency financing projects are as follows: (i) mismatches between the solutions attempted and local institutional environments; (ii) lack of proper balance between and concentration on combining financial intermediation functions and technical assessment functions; and (iii) lack of sustained effort and follow through, especially for adjusting institutional mechanisms and approaches during implementation, in response to market changes or arising operational inefficiencies. Although these points may be well understood by readers from previous chapters, we emphasize that all three of these problems have impacted operational programs in the World Bank and GEF, and require concerted further efforts if the best results are to be achieved in the future. With this in mind, the study team has the following broad suggestions:

- As described in chapter 7, careful diagnostic work on existing in-country financial systems, energy efficiency market conditions, and energy efficiency technical assessment capacities should be undertaken before initiating programs to deliver energy efficiency investments. This work forms the basis for project design. Customized approaches are often required. Projects or programs need not be big or broad to start with—at times it may be best to begin with something small and narrowly focused. But it is useful to maintain a strategic eye toward longer-term energy efficiency development and careful attention to fitting interventions within local institutional contexts.

- For projects involving financial intermediation, it is recommended that parallel attention be given from the outset to (i) the details of developing capacities and mechanisms for financial intermediation aspects and (ii) project pipeline development and technical appraisal. This dual focus should continue throughout program design and implementation. Typically, one aspect may develop faster than the other, and care is required to bring the slower aspect along in parallel.

- It is important to incorporate flexibility in design, so that programs can be adjusted during implementation. Programs should include provisions for serious periodic review, including continuous feedback from clients, and legal documents should be constructed to allow midcourse adjustments to be made relatively easily. Virtually all major programs have required adjustments during implementation, and means to do this efficiently should be considered at the outset.

- All of the above result in exceptionally high labor intensity for program management, operation, and technical support. Major staff and expert inputs are needed to complete diagnostic work, project design and consultation, and development of relatively complex programs. But what is often overlooked is the continued labor intensity required during program implementation, and this has been a specific cause of shortfalls or even failure in some projects. High-quality and concentrated time from program management and expert personnel is essential for new institutional mechanisms to be nurtured along to success.

In summary, energy efficiency financing operations are relatively costly and time-consuming to develop and implement. Development of the associated new institutional mechanisms requires intensive, multiyear efforts. If it is not possible to organize such efforts, it may be best to not attempt such ambitious programs. However, where possible to organize, these programs can make a major, positive difference. With strong returns in terms of financial benefits to enterprises and energy consumers, and with very high potential returns per unit of public investment in environmental and energy security benefits to countries, further development of delivery mechanisms for sustainable energy efficiency financing undoubtedly has a major role to play in meeting the challenges of energy development and climate change abatement.

NOTES

1. As presented in the Three Country Energy Efficiency Project Brazil Country Report, the technical potential for electricity savings in all sectors is estimated to be close to R$5 billion (more than US$2 billion) annually. In the public sector alone, annual savings of US$750 million in electricity costs could be achieved with energy efficiency investments.
2. The potential for energy efficiency investments remains high in India. The India Country Report quotes some previous studies, claiming estimated energy savings in all sectors of more than 50 terawatt-hours with an investment potential of R$140 billion (about US$3 billion).

PART II

ENERGY EFFICIENCY FINANCE CASE STUDIES

INTRODUCTION TO PART II

Part I of this book linked institutional economics concepts with energy efficiency investment planning models and summarized lessons learned from energy efficiency financing operations in general and the Three Country Energy Efficiency Project in particular. Part II presents case studies that either demonstrate applications of models and lessons or chronicle the processes through which the lessons were learned and sometimes both. Some of the cases do a better job of chronicling transitions than of demonstrating current best practice, since the on-the-ground experience in developing and implementing the programs represented by the cases predate and drove much of the knowledge that Part I codifies.

Four concepts stand out from the discussion in Part I and provide four alternative ways of focusing the presentation and discussion of the cases in Part II:

- the three broad "mechanisms" used to deliver energy efficiency finance

- the observable relationship between country institutional contexts and the options for implementing efficiency improvements

- the role of innovations in "deal structuring" (reflected in the flow/organization charts in each case study) that occur at the interface between financing mechanisms and country institutional contexts
- the search for ways to design energy efficiency projects that serve as vehicles not only for reducing current transaction costs faced in making the current deal work but also for shaping the evolution of institutions to facilitate the emergence/development of more responsive markets for energy efficiency transactions

THE FINANCIAL MECHANISMS

As noted in Part I, there are three basic types of institutional delivery mechanisms for energy efficiency investment projects that have been popular in recent years: (i) loan financing and partial loan guarantee schemes; (ii) ESCOs; and (iii) utility DSM programs. It is also common to mix more than one of these in development of specific investment delivery programs.

Analysis of the experience leading up to this book led to the conclusion that the actual relevance and the means of applying versions of these mechanisms varied significantly in different countries and institutional environments, as suggested by energy efficiency program experiences in these three countries.[1]

COUNTRY INSTITUTIONAL CONTEXT

The second item that stands out from Part I is the slowly emerging lesson from investment financing experience that energy efficiency is, at its core, an institutional development problem and that a major factor in the inefficient use of energy is weak financial sectors, which other researchers have demonstrated to arise from weak economic institutions. In particular, weak contracting institutions affect the ability of private actors to design and follow through on agreements regarding present and future ownership and use of resources—in this instance, those related to efficient use of energy. Since these institutional settings are time-dependent and path-dependent, contracting and other economic institutions will differ from one country environment to another and between time periods within those same countries, as institutional development tends to occur over long periods of time.[2] Dynamic or at least evolving institutional environments are likely to be found in emerging market countries that are developing rapidly (such as China, India, and Brazil), in particular where market-based approaches to development promoted by major donor agencies make institutional development a core part of the development assistance packages. The study team concluded that the successful design of energy efficiency interventions must take these differences into account, and consequently the case studies emphasize the differing national settings in terms of contracting and related financial and economic institutions.

ALTERNATIVES FOR STRUCTURING DEALS

The third and in many ways the most critical item in all the case studies is the interplay between the financial intermediation mechanisms and the national institutional environments—that is, differences in approach to "deal structuring" (referred to in places in Part I as "institutional mechanisms" or as "investment delivery mechanisms"). The study team found that, at their core, the projects and programs they were helping to develop were exercises in deal structuring, which involves finding ways of structuring incentives and of managing the risks and maximizing the probability of getting paid (or, on the opposite side of the exchange, getting the product or service that was agreed upon). The issue of getting paid comes out explicitly in Part I in the discussions of ESCOs, the financial intermediation mechanism with forms that have the greatest contract intensity[3] of all the alternatives.

INSTITUTIONAL DEVELOPMENT MECHANISMS

The fourth item that the case studies can demonstrate relates to whether and how the related projects and programs facilitate and support the development and implementation of new or changed institutions, organizations, and procedures that enhance private sector deal-making capacity more broadly (in this case, energy efficiency deals), rather than the immediate investments that are being financed by the project or program at hand. Projects and programs under this heading help create a changed economic environment that reduces the number of deals that cannot be realized because of high transaction costs, while enhancing the scope for private actors to both imagine and structure energy efficiency deals that contribute to overall national well-being.[4]

Deal structuring involves the use of varying combinations of formal and informal institutions and varying combinations of private and public organizations to structure incentives, lower transaction costs, and control the default/performance risks of potential welfare-enhancing transactions. Thus, deal structures within the same country environment will tend to have traits in common with other deals structured in that same environment. By the same token, the same or similar contracting problems will be solved in different ways in different countries. As pointed out by Ménard and Marais (2006, p. 18) in their critique of the IFC (World Bank Group) *Doing Business* studies, "... this method [the *Doing Business* survey] ... endorses a very strong and implicit assumption, which is that in all countries, the same legal patterns are used to solve the same type of business issues. As lawyers specializing in comparative legal studies have known for centuries, this is not so. The real issue is to identify and compare the many ways through which the same issue is tackled in different legal environments and in different countries."

Nevertheless, the study team finally concluded that what matters most in organizing and presenting the case studies is that they be sufficiently diverse to demonstrate many different applications within the above four dimensions. In terms of practicality, the least ambiguous of the four dimensions turns out to be the first one—financing mechanisms. Thus, the following cases are presented in groupings according to the mechanisms that the various country programs were attempting to implement.

NOTES

1. As a result, case studies from countries beyond the Three Country Energy Efficiency Project core of Brazil, China, and India have been included in order to observe variations across a more divergent range of country contexts.
2. Since institutional development occurs over long periods of time, some law and economics researchers find advantages for financial sector development and for the development of contracting forms residing in the evolutionary flexibility posed by the British common law tradition versus civil law traditions. See La Porta et al. (1998, 1999).
3. Fernandez and Kray (2005) discuss issues raised by contract intensity in the context of India.
4. The larger goal of creating an institutional environment that promotes an efficient and equitable financial sector and private contracting institutions in general is well beyond the scope of energy sector operations, although we have maintained throughout this volume that each sector's programs and projects should, at worst, "do no harm" to that broader purpose. Nevertheless, well-conceived energy sector activities will make at least marginal contributions toward broader institutional development goals.

1. CHINA ESCO LOAN GUARANTEE PROGRAM

INTRODUCTION

This case study describes a major national Loan Guarantee Program launched in late 2003 to assist ESCOs operating in China in obtaining commercial loans from local banks. It is different from other major guarantee operations supported by the GEF in that its objective is solely to support the development of the local ESCO industry, and not broader energy efficiency investments. The program is operated by China's largest guarantee company, using GEF resources held by China's government as reserve resources to back the loan guarantees issued.

The program was designed to fit a partially reformed financial sector as best as circumstances could allow. This guarantee program was one of the first of what are now several guarantee programs developed by the World Bank using GEF financing to catalyze additional lending for energy efficiency projects.

PROGRAM ORIGINS AND OBJECTIVES

The China ESCO (or EMC) Loan Guarantee Program is supported under the World Bank and GEF's Second China Energy Conservation Project, approved in 2003 with a total of US$26 million in GEF grant financing. The "full-service shared savings" ESCO concept introduced and operated with support from the first China Energy Conservation Project proved quite successful (see Case Study 7), and new ESCO companies began to emerge around the turn of the millennium, copying the performance contracting ideas developed under this first project. The Second China Energy Conservation Project was designed to scale up China's ESCO business—to help create a new and growing ESCO industry nationwide that could play an important role in delivering commercially viable energy efficiency investments.

In the first China Energy Conservation Project, the World Bank provided lines of credit to the three demonstration ESCOs, which provided a firm financing channel that allowed companies to concentrate on adapting energy performance contracting to China's market. The second project aims to bring local financing institutions into the industry as sustainable sources of finance. Accordingly, a Loan Guarantee Program model was selected as the instrument to introduce local banks to the business, and develop them as important players in the three-party ESCO investment mechanism (involving clients, financiers, and ESCOs), see figure CS1.1. In addition, the Second Energy Conservation Project supported the creation of the EMC Association of China (EMCA) as an institution of ESCO mutual support that would also provide technical assistance to newcomers and would act as a representative of this emerging industry to government and other parties.

The ESCO Loan Guarantee Program provides partial guarantees for loans involving ESCOs and only investments in energy performance contracting. The Second Energy Conservation Project has specific annual targets for achievement of energy performance contracting investments by China's growing ESCO industry, as agreed between the World Bank, GEF, and the Government.

INSTITUTIONAL ARRANGEMENTS: THE FLOW OF FUNDS

The Guarantee Program is currently operated by I&G—although provisions exist to add additional guarantee entities, if necessary. I&G operates the program according to a Guarantee Program Implementation Agreement, agreed between I&G and the Government. This Implementation Agreement establishes all of the essential operating arrangements and the responsibilities of the parties.

GEF funds of US$22 million are allocated in the project for disbursement in installments by the World Bank into an ESCO (EMC) Guarantee Program Special Fund account, held by China's Ministry of Finance. Although the GEF resources remain state assets held by the Government, I&G is able to issue irrevocable loan guarantee

Figure CS1.1. Structural Overview of the EMC Loan Guarantee Program

Source: Authors.

agreements, in I&G's name, but backed by the Special Fund, based on the Implementation Agreement. In addition to the initial deposits of GEF funds, inflows to the Special Fund include interest earned on the account (or instruments held in its name), all guarantee fees charged by I&G under the Program, and recoveries from any subrogated guarantees. Outflows from the Special Fund include payment of I&G's management fee and subrogation payments. I&G's management fee is set as a percentage of the Program EMC loan guarantees it issues, adjusted for actual levels of defaults and subrogation recoveries achieved relative to agreed targets.

I&G operates the ESCO Loan Guarantee Program according to the standard guarantee industry practices of China, with some adjustments to meet the specific goals of the Second Energy Conservation Project. China's loan guarantee industry, operated by I&G and a number of other guarantee companies, developed in an environment (which existed until mid-2005) where loan interest rates were fixed within a very narrow band, and hence, the loan pricing of banks could incorporate little adjustment for risk. In this environment, loan guarantee companies were in a position to support projects that banks considered too risky for the standard loan pricing, and charge customers additional guarantee fees for this service. The business emerged, however, with the guarantee companies providing 100 percent loan guarantees to the banks—in essence shouldering all of the risk and using the banks as agents for loan processing. With the aim of gradually developing commercial lending to the EMCs by the banking sector, the Second Energy Conservation Project has sought a better risk-sharing arrangement between the guarantee companies and the banks. Attempting to balance the traditional practices with the project objectives, loan guarantees have been set to cover up to 90 percent of loan amounts, but with goals to reduce this during program implementation.

Although they are entirely separate organizations, I&G and EMCA operate in close cooperation, as envisaged during project design. While I&G maintains some additional business development channels, EMCA plays an essential role in the development of the market and identification of potential customers for I&G's program. EMCA's role consists of introducing the program to its ESCO members as an important potential source of support in arranging credit, integrating I&G staff into a variety of training and mutual support events for members, and making project proposal referrals.

I&G developed initial systems to complete loan guarantee project assessment and appraisal functions prior to the approval of the World Bank/GEF project, and then put these systematically into operation early during project implementation. I&G created a small, special unit of full-time staff for operation of the project, but this unit also relies on other departments in the company to complete its work—including the company's appraisal department. I&G developed credit assessment and appraisal methodologies specifically for the ESCO business, with some technical assistance provided by the World Bank. These methods rely heavily on the company's well-established existing techniques and appraisal procedures, but add new, specific project appraisal systems to include assessment of the energy efficiency projects in host enterprises to be covered in the guarantees and assessment of the financial standing of the ESCOs themselves. With no special prior experience in evaluating energy efficiency projects, I&G engaged a technical advisor to work full time on technical review, and maintains contact with a small pool of individual, outside technical experts that it can call upon when needed. I&G also utilizes the informal technical advice of EMCA staff and others associated with the World Bank project.

RESULTS OF THE PROGRAM TO DATE

Over three years of implementation through 2006, I&G has issued a total of 85 ESCO loan guarantees, totaling Y 247 million (US$32.1 million). These guarantees have mobilized commercial loans from 11 different Chinese banks totaling Y 274 million (US$35.6 million) in support of EPC investment of Y 440 million (US$57.2 million). Loan guarantee support has been provided to 29 different ESCOs, which represents almost half of the total number of companies who have been confirmed as implementing EPC projects in the country in 2006. These ESCOs are located in 16 different provinces.

As also discussed in Case Study 7, China's ESCO industry grew rapidly during 2003–06, achieving increasingly strong commercial success. Total energy efficiency investment using EPCs in China reached about US$250 million in 2005 and US$280 million in 2006, generating energy savings of 18 and 21 million toe, respectively, over the lifetime of the project assets generated. The ESCO industry investment and energy savings goals of China Second Energy Conservation Project during these initial years have been exceeded by several orders of magnitude. The increasingly strong push of the Chinese Government to increase demonstrable energy efficiency results through all types of measures, market dynamics, increasingly widespread comfort with EPC mechanisms, the strong emergence of EMCA, and the ESCO Loan Guarantee Program have all contributed to this success.

Interestingly, the role of loan guarantees from this program in total ESCO investment has been far lower than originally anticipated. Investments backed with loans guaranteed by I&G accounted for 9 percent of the total during 2005–06. Financing through arrangements with shareholders or other strategic partners has played a strong role in investment. Relatively mature ESCOs are often able to arrange financing from partners or financial institutions without needing to pay for the program's loan guarantees. As a result, allocated funds for the program's Reserve Fund have yet to be fully disbursed from GEF resources.

However, the impact of the Loan Guarantee Program on the ESCO industry is larger than the volumes of direct loan guarantee commitments achieved so far might suggest. Many of the ESCOs that have entered the market had little real understanding or experience with commercial finance, and began their operations with equity finance and resources obtained through shareholder connections. Often, I&G was their first real point of contact with any formal financial institution, and I&G has introduced them to the requirements and processes of commercial loan financing. Many ESCOs obtained their first commercial loan through the I&G program, and then, as they matured, the ESCOs utilized the guarantee program more selectively while obtaining much of their required finance through other channels.

I&G is currently making efforts to expand the types of products it offers in an effort to increase market penetration. New products being introduced in 2007 include guaranteed lines of credit, various types of mechanisms to leverage cash flow from previous ESCO investments, and products to more directly back guarantees of energy efficiency performance.

SUMMARY OF THE ADVANTAGES AND DISADVANTAGES OF THIS APPROACH

At the end of 2006, the China ESCO Loan Guarantee Program had progressed through only about one half of the implementation period planned under the Second Energy Conservation Project, and was still evolving. Hence, final conclusions on the advantages and disadvantages of this approach need to await the results of further evolution and implementation. At the program midpoint, perhaps two points may be offered:

Use of a local guarantee company for implementation. Most other energy efficiency loan guarantee programs have relied on an international entity as the direct underwriter of the guarantees, working in cooperation with local banks. In this project, a local guarantee company was assigned to implement all aspects of the program, utilizing funds disbursed up front to the Government as backing. Assignment of the oldest, largest, and strongest guarantee company in China as implementing agency had a strong advantage in that it brought quick traction for implementation. I&G has a high degree of professionalism and experience in the guarantee business in China, and brought it to bear on the project. I&G also brought instant credibility to the program, and the soundness of its financial guarantee has never been questioned in the market. I&G has been able to work through many local, smaller guarantee companies, with which it has longstanding connections. On the other hand, the longstanding business model of I&G and most other Chinese guarantee companies is to guarantee all of the credit risk of loans—in essence to undertake much of the basic loan appraisal and risk mitigation functions usually undertaken by banks, with banks then playing more of an agent, processing, and collection role. As result, the involvement of the banks in appraisal and risk mitigation is less active than in some other models, even though the program has insisted that the banks assume some share of credit risk. The uptake of ESCO loan businesses by banks themselves as a result of the program has been slow.

Combination of public institutions and commercial interests. The project design involves a complex mixture of public goals and institutions with commercial goals and institutions, which has worked but is cumbersome and difficult for the project implementing agencies. Program implementation involves (i) three large public institutions—the World Bank (policy and implementation), the National Development and Reform Commission (NDRC, policy and implementation), and the Ministry of Finance (implementation and fiduciary responsibility for the Reserve Fund); (ii) a state-owned guarantee company charged with meeting public objectives in a commercial way; and (iii) commercially focused banks and program beneficiaries. The different perceptions and objectives of the different institutions involved create a difficult balancing act for the I&G implementation team. The program is less nimble and flexible than what might be achieved through a purely commercial setup. The public goal of the project—to develop and support one specific, small portion of the energy efficiency industry—also creates major market constraints on the program.

2. HUNGARY ENERGY EFFICIENCY GUARANTEE FUND[1]

INTRODUCTION

IFC, the private sector arm of the World Bank Group, and GEF financed the Hungary Energy Efficiency Co-Financing Program (HEECP) in partnership with local financial institutions (FIs) to build a sustainable commercial lending business in Hungary for energy efficiency investment across a range of sectors. The Program supports development of the Hungarian energy efficiency lending market through the establishment of specialized financial products and the building of new capabilities among Hungarian FIs and project developers to undertake energy efficiency investments. The two primary tools used by IFC include a partial risk guarantee provided to the FIs and a portfolio of TA support for FIs and project developers. It was the first time that a guarantee instrument was used to facilitate commercial energy efficiency lending, which has since been further refined and mainstreamed by IFC and used in different ways in World Bank/GEF projects in other countries to support commercial energy efficiency financing (see Case Study 1 on China).[2] The project also is testimony to the intense efforts required to achieve results, both direct and indirect.

INSTITUTIONAL ARRANGEMENTS AND RESULTS

IFC pioneered the use of guarantee mechanisms with the HEECP (see figure CS2.1), using GEF financing and its own funds. Under the program, Guarantee Facility Agreements (GFAs) for energy efficiency transactions are executed with domestic FIs. Subsequently, Transaction Guarantee Agreements (TGAs) are executed individually for each transaction as it is originated. Each TGA establishes a Transaction Liability Limit for the financed project. Eligible transactions of these FIs are then covered by a partial guarantee, for which IFC charges a guarantee fee. In case of a default, the payment is to be released immediately by IFC to the FI, which then needs to begin the recovery

ENERGY EFFICIENCY FINANCE CASE STUDIES 171

Figure CS2.1. Hungary Energy Efficiency Co-financing Program Institutional Arrangements

```
Guarantee program structure

    IFC  --Investment $-->  IFC/GEF  <--Grant $--  GEF
                                │
                    Guarantee Facility Agreement
                                │
                                ▼
    Transaction            LOCAL                Technical
    guarantees         FINANCIAL                assistance
                       INSTITUTION
                                │
                         EE project loans
                                │
         Vendor         End user          ESCO
          │                                 │
         Lease     Energy services          │
                    agreements              │
        End user                         End user
```

Source: IFC.

process. With one participating bank, a retail guarantee was developed that targeted individual homeowners in block housing and was structured on a portfolio basis, funding a loss reserve.[3]

Under the GFAs, the FIs are responsible for originating and structuring transactions as well as performing the appropriate due diligence and credit analysis. They are also responsible for managing all transactions after closing and pursuing all collection remedies, in the event of default by the borrower. Because only partial guarantees are used and project financing derives from the FIs' own funds, the FIs have clear incentives to originate sound transactions and pursue all collection remedies.

The guarantees were originally targeted at a variety of clients of the FIs. It turned out, however, that the preferred borrowers are project developers: ESCOs, leasing companies, and SMEs that are involved in delivering energy efficiency equipment, projects, and services. By using the guarantees for developing new financial products in these niche sectors, FIs are able to avoid the transaction costs of having to deal with a multitude of small projects proposed by individual borrowers. This also eases the problem of collateral and of securing the savings ("negative cash") stream associated with energy efficiency projects by using energy supply agreements between project developers and end users or performance guarantees provided by the ESCOs. Similarly, TA can be provided more effectively to a small number of project developers.

TA is provided to the participating FIs to train bank staff (particularly credit officers in appraising energy efficiency projects), and for ESCO capacity development—for example, developing projects with ESCOs and providing emerging ESCOs with business planning to assist in capitalization, business development, and joint venture partnerships.

The project is administered through the local IFC office with support from a Supervisory Committee composed of members of different IFC departments. The Supervisory Committee must approve the selection of FIs, the granting of each transaction guarantee, approval of credit criteria for portfolio guarantees, and all general policy questions. Management tasks include the development of relationships with FIs, review of energy efficiency project financing transactions, negotiation and documentation of agreements, provision of TA, and so forth.

HEECP consists of three different phases, each of them characterized by different parameters, summarized in table CS2.1. Two FIs participated actively in the pilot phase HEECP1, which received US$5 million of GEF financing. After the positive experience with the HEECP1 pilot, IFC contributed an additional US$12 million to a second phase of the HEECP, and an additional US$0.7 million of GEF funds was provided for TA purposes. Additional TA funds of US$0.35 million also were granted by IFC's Austrian and Netherlands Trust-Funds. Six FIs, making up more than 50 percent of the Hungarian FI market in terms of assets, were involved in this phase. Based on the new GFAs, FIs need to appoint two senior managers, responsible for credit and for marketing and origination, respectively, to oversee the FI's participation in the guarantee program. This requirement is intended to assure that the guarantee is recognized in credit committee decision processes as a credit risk management tool, and that the

ENERGY EFFICIENCY FINANCE CASE STUDIES 173

Table CS2.1. Evolution of HEECP Parameters, 1997–2006

Parameter	HEECP1 (start 1997)	HEECP2 (start 2001)	HEECP3 (merged with CEEF Program 2005)
GEF Contribution	US$5 million, of which US$0.75 million for TA and administration	US$0.7 million for TA purposes	n.a.
IFC Contribution	US$0.3 million for management and operation	US$12 million for guarantee and US$0.5 for management/operations, and IFC's Trust Funds US$350,000 for TA	n.a.
FI Contribution	Match general support funds. Self-funded loans	Match general support funds. Self-funded loans	Match general support funds. Self-funded loans
Number of participating FIs	3 (2 actively)	6 (95% of the Hungarian energy efficiency lending market)	4
Guarantee type	50%, subordinated recovery	35%, subordinated recovery	50%, pari passu
Additional guarantee product	First loss reserve portfolio product (GEF)	None	Portfolio-based (up to 5%) first loss reserve (GEF)
Max. guarantee, %	50	35	50
Max guarantee size	US$0.5 million	US$0.5 million	US$2 million
Guarantee fee	1%	1%	Varies according to market
IFC:GEF ratio	0	Initially 2, up to 3	Initially 2, up to 3.58
Availability	1997–2000	2001–2004	2005–08 (linked to the life of CEEF)
Potential for streamlined decision making	Low, only for small exposures, since risk sharing is not symmetrical	Low, only for small exposures, since risk sharing is not symmetrical	High, with symmetrical risk sharing
Eligible projects	Excludes nonprivate entities	Excludes nonprivate entities	Includes municipally owned entities with independent corporate governance
Currency of guarantee	US$	US$	Euro

n.a. = not applicable. CEEF = Commercializing Energy Efficiency Finance.
Source: Based on Sources quoted in note 1.

guarantee product and energy efficiency finance products are disseminated throughout the FI. The maximum guarantee is reduced from 50 percent to 35 percent, and the guarantee fee is slightly increased to more closely reflect market rates.

By the end of 2006, the US$55 million loan portfolio that HEECP directly supported with US$17 million of guarantees represented US$93 million worth of energy efficiency investments with a total of 331 energy efficiency projects and 1,500 contracts in the gas retail portfolio. No guarantee has been called to date; and only the gas retail portfolio suffered defaults of US$0.15 million. Figure CS2.2 shows that after a relatively long period of modest activity as they gained familiarity with the new instrument, banks significantly scaled up their energy efficiency lending with support of the guarantee program in 2004.

In addition, HEECP provided TA advisory support to more than 30 transactions and for development and establishment of five specialized financial products. These products have yielded a substantial sustained pipeline of new investment supported by HEECP's partner FIs (visible in figure CS2.2), with only a portion of the transactions utilizing the

Figure CS2.2. HEECP Results, 1997–2006

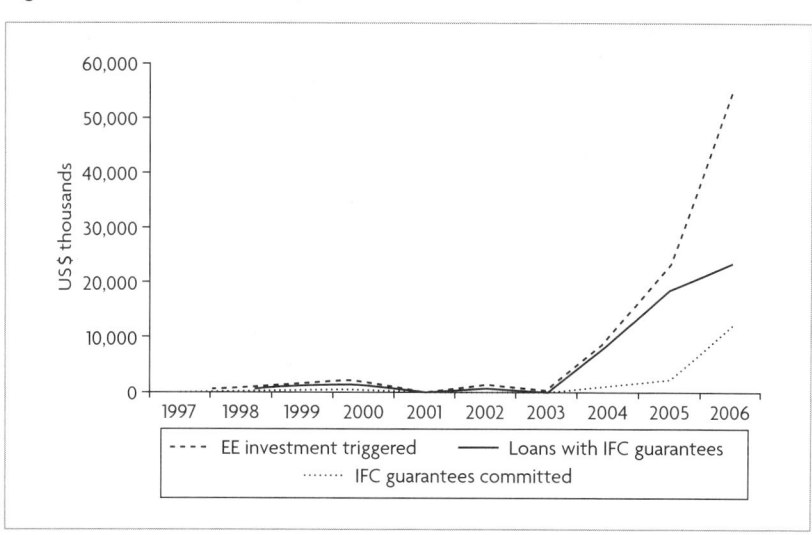

Source: IFC.

guarantee. The above-mentioned types of projects are increasingly being financed without the IFC guarantee. Banks are hunting for energy efficiency projects on their own, are requesting lower levels of collateral and down-payments as they become increasingly familiar with the risks of such projects, and are, at times, financing projects based on cash flow alone.

This is a substantial change from program pilot phase HEECP1, when the maximum guarantee was 50 percent, and security packages were initially fairly standard, typically including ownership of the leased assets, a preferential drawing right on the project developer's company bank account, end user guarantees, and preferential drawing rights on the bank account of the end user, in addition to the IFC guarantee.

The original HEECP project was extended to 2008 within the framework of IFC's Commercializing Energy Efficiency Finance (CEEF) project (see note 2) with the objective to transfer lessons learned, extend the program into additional significant markets, and refine the financial products initially used based on experience. Modification of the HEECP guarantee product, to harmonize it with the CEEF financial product, includes the introduction of a pari passu-based guarantee (shared equally between the parties, replacing the subordinated recovery) and the limited use (up to 5 percent) of a first-loss portfolio-based guarantee. Among the new markets and products of HEECP3 are blockhouse portfolios, an SME facility and some programs for renewable energy, reflecting the need for specialized financial products that address still undeveloped niche markets.

A further modification in HEECP3 includes a new credit approval system to encourage FIs to develop standardized energy efficiency products, and build loan portfolios based on these products, backed by the Program's guarantees. The first standardized product has been developed for Raiffeisen Bank in Hungary. Under the framework of their blockhouse energy efficiency finance product, the local IFC country manager approves transactions and issues transaction guarantee agreements on behalf of IFC based on pre-approved criteria.

HEECP's overall contribution to the development of the energy efficiency finance market in Hungary goes well beyond the direct influence of approved guarantees. The Program facilitated the expansion of the energy efficiency market to new end users, and the development of new technologies and new financial products. The Program has brought considerable knowledge and experience for the development of new financial products at FIs where the market opportunities otherwise would be ignored and lending opportunities automatically rejected. Particularly successful are new lending approaches for blockhouse renovation, cogeneration, and street-lighting projects.

The IFC Program has benefited from several market and institutional changes, including the following:

- lower interest rates and stronger competition in the banking sector
- increasing energy prices
- changes in government policies helping the development of new markets such as cogeneration, which benefited from off-take agreements for electricity at attractive rates
- legal changes that facilitated the formation of housing associations, which could then pledge common assets as loan security
- existence of generous subsidies for renovation of residential buildings and heating systems

THE ROLE OF PROJECT DEVELOPERS AND THEIR RELATIONSHIP WITH THE BANKS

Most loans under the HEECP are provided to ESCOs and other project developers such as manufacturers or installers of energy efficiency technologies. Packaging of energy efficiency projects by these developers has helped reduce transaction costs. Hungary now has a very well-established ESCO industry that is quite familiar with energy performance contracts (EPCs).[4] Most ESCOs are small and provide relatively straightforward energy efficiency solutions, for street-lighting or boiler replacement, for example. About 10 ESCOs, including several big international players, are able to develop and implement more complex projects under EPCs, such as cogeneration solutions. Some banks have even set up their own ESCOs.

As for other clients, the balance sheets, technical strength, and projects of ESCOs are assessed by the banks as part of their internal procedures. If necessary, outside experts, who often have ongoing relationships with the banks, are invited to provide their technical assessments. For unfamiliar projects/technologies, loan security requirements initially have been very strict (for example, requirement of 40 percent self-contribution to investment compared to a more normal 10 percent) to cover engineering and other risks. Once the banks understand a particular business, they tend to revert to normal security requirements.

THE IMPORTANCE OF TECHNICAL ASSISTANCE

TA, including both funded initiatives and more informal contacts between IFC staff and market players, was crucial for the eventual success of HEECP. TA was provided to banks, project developers, and other energy efficiency market participants and can be divided into the following categories:

- At the beginning of the program, TA was important in marketing the idea of energy efficiency and building general awareness about HEECP and its benefits among the banks, potential project developers, and end users. During this phase training was provided to the staff of participating banks on the energy efficiency market and on how to work with IFC. Energy audits of some facilities were concluded by local experts, but later discontinued when they did not contribute to a substantial deal flow.

- External experts (international consultants) supported program partners to find solutions for turning market opportunities into bankable solutions. Most importantly this consisted of financial structuring and product development, resulting in several specialized energy efficiency products for the participating banks. Examples included products for gas heating, blockhouse renovation, blockhouse heating, streetlighting, heat modernization and district heating renovation, and cogeneration. New products were usually marketed under the lead of the FIs, with HEECP occasionally providing cost-sharing. In some cases ESCOs and other energy efficiency businesses received financial advisory support for business planning, corporate finance planning, and raising equity.

- TA in the form of contacts between IFC staff and program partners became increasingly important. This includes discussion of ideas for new financial products and schemes. An example is the recent portfolio guarantee incorporated into HEECP3. If necessary, FIs will subsequently call in experts for more detailed engineering analysis or support in structuring the transaction, which HEECP may cost-share.

ADVANTAGES AND DISADVANTAGES OF GUARANTEE PRODUCTS

The Hungary project has shown that the commercial financial sector can provide increased lending for energy efficiency investments if the project puts in place the following elements:

- initial champions to prove that energy efficiency projects are financially viable and that financing can be structured with the savings stream providing security for lenders
- credit enhancements to substitute for straight guarantees and collateral
- development of niche financial products that complement the existing business and expand the FI's client base
- project development activities and capacity building, supported from TA grants

The primary objectives of the HEECP were the reduction of credit risk and reduction of transaction costs in energy efficiency financing. Where financial markets are fairly well-developed as they were in Hungary, a guarantee mechanism can be a promising approach to achieve these objectives. In other places, different methods may be needed to increase financing through the banking sector.

NECESSARY CONDITIONS AND LESSONS LEARNED

The experience with HEECP shows that the appropriate markets for a loan guarantee mechanism need to have adequate liquidity, attractive interest rates, competition, and reasonably mature financial institutions that are willing to accept some risks. The guarantees together with TA can then successfully mobilize existing resources. It is important to emphasize that guarantees alone cannot solve systemic banking or credit problems.

Key among the lessons is IFC's experience with the *sustainability* of the guarantee approach. IFC undertook HEECP with the expectation that the guarantees would no longer be required by FIs after the program life in order for energy efficiency projects to continue to be financed. This basic sustainability had already been achieved with the HEECP1 pilot. FIs have now adopted the guarantee as a surgical instrument, which they use on a strategic basis to support early development of new sector markets and first-of-a-kind transactions. The participating FIs quickly gain comfort with the new financial products and the risk profiles of the new energy efficiency deal types, and then continue lending without use of the guarantee (and without paying guarantee fees). While the absolute volume of energy efficiency lending is comparable to what was projected, the absolute volume of guarantees required by the FIs in building an energy efficiency lending business was lower than anticipated.

HEECP experienced a slower start-up than initially hoped which would also be expected for most other introductions of new financial products for energy efficiency in other settings. Six months or more was needed for the development of new products with FIs. The lead time for full utilization of the new products was even longer; to generate significant volumes of deals took at least 1.5–2 years of product life. Since these were pilot operations, there was no experience from similar programs before implementation. Another key lesson is the need to invest substantial time and effort to build energy efficiency lending businesses in FIs where there was no prior experience, and the critical importance of TA support for building that FI capacity. The learning curve was longer, and building capacity and creating demand for the guarantee product required more up-front investment from the program than was expected. TA was also invaluable for developing special products such as the loan loss reserve fund for portfolios of small standard loans, which require the development of standardized loan applications and appraisal methods.

NOTES

1. Unless otherwise noted, information in this case study is based on http://www.gefweb.org/Documents/Project_Proposals_for_Endorsem/HEECP_and_CEEP_Amendment_Document.pdf, "Supplemental Evaluation of HEECP2," by the Danish Energy Management, Report for IFC, May 2005, and personal communications with IFC staff.
2. IFC: CEEF Program in the Czech Republic, Estonia, Latvia, Lithuania, Slovak Republic, and Hungary; in combination with an IFC credit line in Russia. World Bank: China, Croatia, Poland, Bulgaria, the Philippines, Tunisia, Algeria, Macedonia.
3. A Retail Gas Program has been undertaken with Raiffeisen Leasing to provide lease financing in partnership with a local gas utility for individual homeowners to install new gas boilers and a range of building envelope improvements. A loss reserve was established to provide credit support for a portfolio of project leases. A first portfolio has been closed with 2,050 leases and total principal of US$1.5 million. Losses—that is, the principal value of leases for which payments are overdue by three months or more—have been 4.4 percent of the total value of the portfolio.
4. See, for example, Ürge-Vorsatz, Langlois, and Rezessy (2004).

3. ROMANIA ENERGY EFFICIENCY FUND

INTRODUCTION

By the end of the 1990s, Romania still had one of the highest energy intensities (and greenhouse gas intensities) in Eastern Europe and—at the same time—a history of failed energy efficiency investment initiatives, including energy efficiency credit lines and ESCO promotion schemes. Market surveys showed that the demand for investments in more efficient equipment was huge, but that internal funds were the only financing source being utilized. The financial sector was characterized by dominant public sector borrowing, very high interest rates, concentration on short-term lending for working capital, and exorbitant collateral requirements. In this environment, the World Bank and GEF decided to support a free-standing, revolving energy efficiency loan fund: the Romanian Energy Efficiency Fund, or FREE (Fundul Roman pentru Eficienta Energiei). The fund would be led by a professional fund manager (FM) and would combine project development, technical-financial and creditworthiness evaluation, and financing at commercial rates. It was hoped that by showing that such a fund would grow, and that spreads and fees from the loan portfolio could cover the costs of running the fund, thus ensuring sustainability, banks would become interested in cofinancing projects and eventually begin to finance energy efficiency projects on their own account.

However, FREE was barely operational when financial market conditions in Romania changed dramatically (see table CS3.1). The government could now borrow more cheaply on international capital markets, dollar-denominated interest rates declined from close to 20 percent to under 10 percent,[1] and commercial banks began to compete for new clients. Under these new competitive conditions, FREE has had a very slow start-up in view of its original business plan. Since lending now takes place at interest rates in the 4–6 percent range, FREE's revenue is reduced to levels that barely cover its operational costs. Although its business has shown an upward trend, FREE

Table CS3.1. Romania Financial Market Conditions

Indicator	2000	2006
Inflation	40.7%	4.9%
Interest rates		
• LIBOR (1 year)	6.66% (Jan.)	5.49% (Jan. 07)
• NBR policy rate	35% (Jan. 02)	9%
• Average rates for RON-nominated loans to nongovernmental, nonbank clients	42% (Jan. 02)	13% (Jan.)
Financial intermediation (nongovernmental credit/GDP)		
• Romania	9.3%	21.1% (2005)
• Average of 10 Non-EU member states	n.a.	40.4%
Total credit in billion RON (real)	10 (Jan.)	23 (July)
Credit to corporate entities (billion RON [real])	5.5	18
• Short	4	9
• Medium	1	5
• Long	0.5	4
Bank loans granted to clients as percent of total attracted and borrowed sources	36% (Dec.)	67% (Nov.)

Source: National Bank of Romania (NBR).
RON = Romanian lei.
n.a. = Not available.

has made only 16 loans, totaling US$6.3 million for investments of US$9.4 million at the end of 2006, far less than expected at project appraisal (see table CS3.2).[2] FREE employs largely traditional lending schemes, requiring collateral rather than recognizing the improved cash flow through energy bill savings which has limited the types of projects it can finance. FREE has only recently secured two cofinancing deals from commercial banks.

This project provides an example of the need to tailor the financing mechanism carefully to the country institutional environment, and more important, to provide flexibility in the products and terms to allow the mechanism to adjust quickly to changing market characteristics. The slow development of the project pipeline also reinforces the need to build up a strong project development function up front.

Table CS3.2. FREE Project Results (US$ Millions)

Project results	2003	2004	2005	2006	Total
Number of projects financed	0	2	6	8	16
Lending volume	0	0.555	2.589	3.194	6.338
Investment volume	0	1.468	3.279	4.642	9.389
Commercial cofinancing	0	0	0	1.870	1.870
Other cofinancing (clients)	0	0.913	0.69	1.448	3.051

Source: Authors.

INSTITUTIONAL ARRANGEMENTS

FREE was established in 2001 through an Emergency Ordinance by the Government of Romania as an independent, autonomous legal entity. Its organizational structure is represented in figure CS3.1. FREE is independent and separate from any government agency, even though the funding initially comes mostly from GEF and is considered public funding. The fund is overseen by a Board of Administration (BoA), consisting of seven representatives from the Romanian private and public sectors, with a private sector majority and an annually rotating chairmanship. The Investment Committee (IC) is a subcommittee of the BoA, including two BoA members and three non-BoA members from the banking sector. The IC reviews all financing proposals submitted by the FM and makes its investment recommendations to the BoA for final decision through majority voting. FREE is administered by a small professional management team headed by an executive director whose main responsibility is to provide overall management of the project and serve as the main liaison with the World Bank and the Romanian government during project implementation.

Following a competitive bidding procedure, FREE entered into a performance contract (with the variable part of the payment depending on achievement of the investment and loan default targets, and a success fee depending on the fund's increase in net value) with a professional FM to manage the investment aspects of the fund in a commercial

Figure CS3.1. FREE Institutional Arrangements and Funds Flow

Source: See note 2.

manner and as a "one-stop shop" for its clients. Being a one-stop shop involves the following:

- originating the deal flow and identifying investment opportunities
- determining the structure of specific investments (including size and use of funds), performing creditworthiness analysis of potential clients, and performing technical/environmental review and financial analysis of investment projects
- structuring co-investment related to specific deals
- recommending potential investments to the Investment Committee
- monitoring, after Board approval, investment implementation and actual energy and greenhouse gas (GHG) savings
- ensuring that financial targets are met

Initially, FREE was designed as a revolving debt fund to finance a mix of private and public sector clients with loans in the US$100,000 to US$1 million range. Projects payback was targeted at three to four years. At least 50 percent of project benefits are required to come from energy cost savings. Since FREE is a nonbanking financial institution, it is not bound by the strict risk management requirements of the National Bank of Romania and can structure collateral more freely than banks.

FREE was designed to be flexible in terms of product mix and terms, enabling the FM to offer financial products demanded by the evolving market. In order to set FREE apart from other financial service providers, the FM was expected to employ innovative risk mitigation measures for the various types of clients, project, and products; to develop appropriate new financial products for energy efficiency investments; and to enter into active partnerships with commercial financing institutions, leasing companies, and ESCOs. FREE also was expected to assist its clients and partners in project development and financial packaging, and to generate and disseminate information on the benefits and costs of energy efficiency investment and success stories. Technical assistance financing was made available for these tasks. However, the project pipeline development capacities of the FM were not strong, and little attention was given to the development of staff or completion of technical assistance activities in this area. Initially it was hoped that FREE could finance projects prepared under FREE's "sister project", the UNDP/GEF project "Capacity-building for GHG Emissions Reduction through Energy Efficiency in Romania." However, business development for FREE through this channel proved not very successful.

FREE's "one-stop" design requires strong institutional capacity to deliver both financing products which can meet the needs of a rapidly changing market environment and to identify and develop a robust project pipeline. The first three years of operation showed that the combination of the particular FM selected through a bidding procedure, the incentive structure which defined the FM's operation, and the oversight of a BoA that was largely risk-averse, given concerns to not allow the project's GEF resources to be lost, was unable to deliver on this difficult challenge. The FM was unable to develop new concepts and partnerships with other financing entities, and inadequate attention and resources were devoted to project development. In early 2006, the World Bank entered into discussions with the Government and FREE for a restructuring which would take into account the new realities in the Romanian financial sector. This involves reducing operating costs and adapting them to the limited income generation potential. The ineffective FM company has been replaced by a more cost-effective team of local energy efficiency and financial experts. The IC can approve small loans of up to US$300,000 without requiring a vote by the BoA. Cofinancing with commercial banks is now pursued more vigorously, although only two cofinanced projects have resulted so far.

A summary of the advantages and disadvantages of FREE, as it was originally designed, is provided in table CS3.3.

LESSONS LEARNED

- Although often difficult to achieve, institutional designs need to be kept as simple as possible, as it is difficult to align incentives for many different parties.

Table CS3.3. Summary of Advantages and Disadvantages of FREE

Advantages	Disadvantages
• One-stop shop for project development and financing with competences in energy efficiency and finance	• Reliance on a competitively procured performance contract with a Fund Manager to deliver the program's core institutional capacity for the one-stop shop proved to be a risky approach
• Ability to operate without reliance on a banking system that was dysfunctional at project appraisal	• Inability to adjust easily to subsequent positive changes and opportunities in the banking sector
• Use of market-based interest rates, but greater potential flexibility in products and terms, as FREE is not subject to National Bank supervision	• Institutional scheme too complicated with many decision makers with different interests
• Technical assistance financing provided for project development and capacity building	• Insufficient incentives to utilize technical assistance for maximum investment results

Source: Authors.

- Development of institutional capacity to deliver on core functions, involving unique combinations of technical and financial skills, is the key challenge for most energy efficiency projects, defining project success or failure. In this project, it proved risky to rely on procurement of the critical capacity needed through competitive bidding, and management through an inflexible performance contracting arrangement with one firm that was difficult to change. This approach subsequently had to be abandoned.

- Project designs need to incorporate strong and reliable project development functions that can successfully build up project pipelines.

LESSONS FROM THE BULGARIAN ENERGY EFFICIENCY FUND

The Bulgarian Energy Efficiency Fund (BgEEF) has incorporated some of the lessons learned from the experiences of FREE into its design.[3] Its institutional structure is simpler and more flexible and it has the potential to develop a larger menu of financial products.

BgEEF has had a relatively quick start-up. In 2006, its first year of operation, 20 projects with a loan amount of US$1.67 million were approved for financing, for investments totaling about US$2.4 million. This reflects aggressive marketing that is outsourced to a private sector entity, but also more familiarity with energy efficiency financing in Bulgaria, initially through a guarantee facility supported by USAID's Development Credit Authority. An EBRD credit-line project (see http://www.beerecl.com) has further primed the energy efficiency investment market. However, this credit-line project provides a substantial subsidy, leaving BgEEF at a disadvantage to attract customers. As a result, BgEEF's projects so far are mostly small and concentrated in the municipal sector and consist only of straightforward loans, provided entirely by BgEEF. The FM team is now working on developing cofinancing framework agreements and on finding workable structures for applications of a guarantee instrument. This project includes no formal structure for project development; the FM relies on clients coming to the fund and on proactive promotion by the Energy Efficiency Agency and several NGOs.

NOTES

1. Between 2000 and 2003, LIBOR decreased from above 7 percent to slightly over 1 percent. At the same time the Bucharest Bank Offered Rate decreased from 70 percent to under 10 percent.
2. http://www.gefonline.org/ProjectDocs/Climate%20Change/Romania-Energy%20Efficient%20Project/PAD-P068062-toc.pdf and http://www.free.org.ro.
3. http://www.gefonline.org/ProjectDocs/Climate%20Change/Bulgaria%20Energy%20Efficiency%20Project/Project%20Brief%20BUL%20BEEF%20Work%20Program%20Incl.%20090404.pdf. See also http://www.bgeef.com/display.aspx.

4. IREDA ENERGY EFFICIENCY LOAN FUND

INTRODUCTION

The Indian Renewable Energy Development Agency Ltd (IREDA), an ISO 9001:2000 certified entity, was incorporated as a Public Limited Government Company in 1987 under the administrative control of the Ministry of New and Renewable Energy (MNRE), Government of India, with the following mission:

> Be a pioneering, participant friendly and competitive institution for financing and promoting self-sustaining investment in energy generation from Renewable Sources, Energy Efficiency and Environmental Technologies for sustainable development.[1]

To achieve this mission, IREDA is operating a revolving fund for development and deployment of new and renewable sources of energy and for energy efficiency investments providing financial support to specific projects. IREDA is one of the largest DFIs specializing in renewable energy and energy efficiency in the world. Since its inception, IREDA has played a critical role in opening the markets and in scaling up implementation of renewable energy and energy efficiency projects in India.[2] This case study shows that high levels of investment can be achieved through the operation of a large, specialized, parastatal financial institution, but also presents the limitations of such an approach once the market gains comfort and experience in lending for energy efficiency.

PROJECT DESCRIPTION AND INITIAL RESULTS

Building upon the initial success of IREDA in lending for renewable energy, the World Bank extended a line of credit and GEF grant to IREDA for developing and financing energy efficiency and conservation projects (in addition to the continuation of financial support for small hydro projects) in 2001 through the Second Renewable Energy Project. IBRD and IDA funds were onlent to private entities at commercial rates (see figure CS4.1 for the institutional arrangements).

Figure CS4.1. IREDA Institutional Arrangements

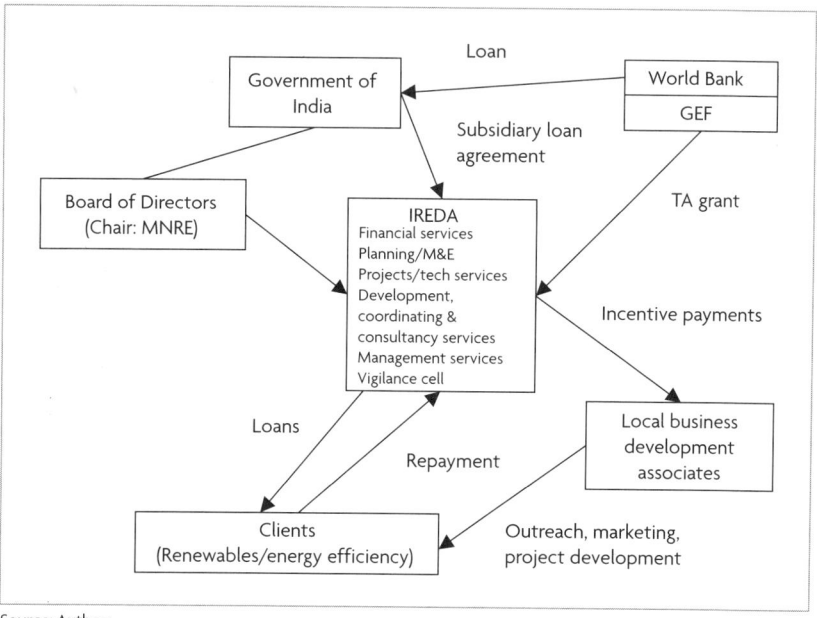

Source: Authors.

The GEF grant component aimed to establish delivery mechanisms for energy efficiency services and equipment, to support implementation of DSM schemes, and to support the development of ESCOs in India through technical assistance and capacity-building efforts.

To operate efficiently as a centralized organization in a large country such as India, IREDA has built up in-house technical and financial expertise, but also relies on some outsourcing. To assist in pipeline development, IREDA has built a network of business development centers and strategic allies throughout India, consisting of about 50 technical consultancy organizations, state nodal renewable energy agencies, local and national productivity councils, private consultancy organizations, NGOs, and technical institutions and agencies registered under the Societies Act. IREDA provides them with training and some financial resources. In addition, they receive incentive payments upon loan disbursement and commissioning. Although certain aspects of the energy efficiency business were new to IREDA's project and technical services department, many of the issues involved in appraising energy efficiency and renewable energy projects are similar. Technical assessment of projects is done in house, although specialized consultancies are used on a case-by-case basis for certain larger projects.

IREDA has already established credibility in the energy efficiency financing market, building upon its successful track record in lending for renewable energy projects. To date it has successfully financed energy efficiency and conservation projects in the following industries: sugar, paper, textile, steel/sponge iron, heavy chemicals, cement, power generation, and DSM in electric utilities, including ESCO projects through performance contracting/revenue sharing. IREDA has sanctioned (that is, obtained approval by its Board) 19 projects to date, totaling US$60 million, of which 11 have been fully commissioned (the loan agreement has been signed, the loan has fully disbursed, and the project has been completed). Average project size is large (US$4 million) as the majority of the current loan portfolio consists of large cogeneration and waste heat recovery projects. However, IREDA has also financed several smaller loans for equipment replacement and DSM (US$200,000–800,000), including motors, control systems, capacitors, lighting, pumping systems, and boilers.

IREDA offers a variety of terms and tenors, covering up to 70 percent of ESCO project costs and up to 80 percent of energy efficiency equipment costs, including soft-cost items such as energy auditing and detailed project report preparation. Loan tenors vary from 6–10 years, and interest rates vary between 8–11.5 percent, which is roughly comparable to loans currently available in the commercial market. A rebate of 1–1.5 percent is provided if borrowers can furnish the security of a bank guarantee. Other loan security measures commonly adopted include mortgage of immovable properties, hypothecation of movable assets, guarantees by promoters, use of trust, and retention or special accounts.

PROS AND CONS OF PUBLIC SECTOR DFI APPROACH

Some of the pros and cons of supporting energy efficiency investments through a parastatal entity such as IREDA are shown in table CS4.1.

Table CS4.1. Pros and Cons of Supporting Energy Efficiency Investments through a Parastatal Entity

Pros	Cons
• Strong combination of in-house technical and financial expertise	• Loan requirements and procedures are seen as too bureaucratic
• Demonstrated financial performance (IREDA is a profit making entity)	• Less flexibility in changing financial products to meet market needs
• Large overall loan portfolio	• Less local presence, branch offices
• Credibility in the market	• Difficulty in sourcing lower-cost funds from the commercial market without special government programs
• Existing network of business development associates	
	• Difficulty in competing with local banks once initial market comfort level is achieved

Source: Authors.

While IREDA possesses several advantages in providing finance for energy efficiency investments, its structure and operating procedures have produced several inherent disadvantages that have caused IREDA to lose market share to domestic commercial financial institutions as interest rates have fallen, overall liquidity in the Indian financial market has increased, and experience has been gained in the successful implementation of energy efficiency projects (reducing their perceived risk). As a result of these changes, IREDA has seen its overall portfolio shrink. IREDA's loan procedures have been criticized as being too bureaucratic and cumbersome by potential loan recipients, especially SMEs that are seeking smaller loans and have less experience in dealing with financial institutions. IREDA's internal procedures in issuing new loan products are less flexible than those utilized by the domestic financial market participants. While IREDA has been able to increase the variety and competitiveness of its loan products (with differing terms and tenors) and has attempted to streamline its procedures, it cannot match the offerings and flexibility of the broader Indian financial market.

IREDA had good success in developing a large pipeline of potential energy efficiency projects through its network of business development associates and through the implementation of certain GEF-sponsored

TA activities, but the success rate of translating these initially identified projects into actual signed loans was low. Initially, certain energy efficiency projects that were sanctioned and appraised, but had not yet signed loan agreements with IREDA, were "shopped" by project sponsors to other commercial banks—using IREDA's appraisal report to validate the actual investment risk—to obtain financing on more favorable terms. Although this behavior negatively impacted direct financing levels under the World Bank line of credit, it did demonstrate progress in meeting the overall World Bank project development objectives of stimulating energy efficiency markets. IREDA has since changed its procedures, and now only provides the complete appraisal reports to clients upon loan signing.

LOOKING AHEAD

IREDA's main challenge is to properly position itself in the increasingly competitive Indian financial market for energy efficiency products and services. IREDA will continue to work to increase the attractiveness of its loan products by reducing its cost of funds and to improve and streamline internal procedures. The World Bank is supporting IREDA in its efforts through several ongoing consultancies. Energy efficiency lending, especially for larger, capital intensive, technically complex projects, is expected to play an important role in IREDA's future loan portfolio.

NOTES

1. http://www.ireda.in/homepage1.asp?parent_category=1&category=6.
2. IREDA is the largest financier of renewable energy projects in India, with total loan commitments of over US$1.7 billion and cumulative disbursements of over US$910 million. IREDA loans have financed over 2,700 MW of new renewable energy and energy efficiency projects.

5. ENERGY EFFICIENCY CLUSTER LENDING FOR SMEs BY INDIAN BANKS

INTRODUCTION

Cluster lending represents an innovative approach that can be utilized by domestic financial institutions to increase lending for energy efficiency projects in the SME sector.[1] These enterprises, especially those in energy-intensive industries in developing or emerging market economies, typically have unexploited opportunities for substantial energy efficiency gains but usually do not have the in-house capacity to prepare and implement the projects needed to achieve those savings. SMEs also commonly face additional barriers in accessing commercial financing in these economic environments even after cost-effective energy efficiency projects have been identified.

Beyond the problems faced by the SMEs themselves in identifying and preparing energy-saving projects, additional barriers are faced by Indian banks in lending to SMEs, including high (per loan dollar) transaction costs and the increased risk associated with lending to smaller clients. The cluster approach can bring specialized technical support and outreach to smaller enterprises along with follow-up loan provision based on a standardized, replicable model that can result in substantial reductions in transaction costs per loan. With support from the Three Country Energy Efficiency Project, the State Bank of India (SBI) adapted its Project Uptech, a cluster lending approach for technology upgradation in India's SME sector, to incorporate energy efficiency improvement projects. That experience, and subsequent further experience in spreading and broadening that approach in other parts of India's financial sector, is the subject of this case study.

BACKGROUND

Lending to SMEs is a government-set priority for Indian banks because of SMEs' importance in generation of economic growth, employment, and exports. The cluster lending strategy for reaching SMEs predates the Three Country Energy Efficiency Project and was initially begun for reasons other than energy efficiency alone. To service the energy efficiency financing needs of this hard-to-reach class of customers, SBI adapted its cluster lending strategy (Project Uptech), and several banks followed with their own versions of the approach. "Cluster lending" refers to lending operations targeted at certain clusters of industries that are co-located for some economic (or policy) reason.[2] The objectives of cluster lending programs include lending for investments to increase SME competitiveness through technology upgradation, cost reduction (through reduced wastage and increased operational efficiencies), increased productivity, and (in some cases) improved product mix.

As implemented thus far, SME cluster lending in India has focused either upon (i) a specific sector or technology group, or (ii) upon a geographically grouped cluster that includes several industrial categories but concentrates on a few technical interventions as a way of minimizing assessment and appraisal costs.

The Importance and the Problems of the SME Sector in India

Though the SME sector makes a significant contribution to India's industrial GDP and exports and is a significant source of employment in urban, periurban, semiurban, and rural areas,[3] SMEs have fallen behind the benchmarks set by larger industrial firms in terms of productivity, performance, efficiency, and technology. The gap has continued to increase even after the comprehensive economic reform program initiated in the early 1990s. Wide-ranging fiscal incentive programs offered to small-scale industry (SSI) units by government have not succeeded in bridging this growing performance gap between SMEs and larger industrial firms.[4]

Early Cluster Lending Approaches

As the realization grew in the late 1980s that lending programs that worked for larger firms were not succeeding in reaching SMEs, the concept of cluster lending began to gain ground. The objective of this newly emerging approach was to increase lending to SMEs while reducing transaction costs per borrowing enterprise to within reasonable and acceptable limits. Many variations have subsequently been tried, with the overall objective of finding a readily acceptable and workable solution that would make it possible for commercial banks to bring cluster lending into the mainstream of their operations.

Essentially two types of cluster lending programs have been attempted thus far:

- those aimed at upgrading technology and improving overall performance in a holistic fashion (with energy efficiency improvements as an integral component of this improvement without being the sole objective)
- those where energy efficiency improvement is the core lending objective

The Indian financial sector has initiated two significant programs of special relevance to this discussion:

- SBI, India's largest public sector bank, set up a technology upgradation (Project Uptech) program for SMEs across all industrial categories in 1988. This was very different from the previous industry specific programs that were targeted at somewhat larger enterprises.
- With the passage of the Small Industries Development Bank of India (SIDBI) Act, SIDBI was set up in April 1990. In keeping with its charter, SIDBI has emerged as the principal financial institution to promote, finance, and develop the small-scale sector. In recent years, SIDBI has also begun supporting medium-scale enterprises.

In terms of banking infrastructure and arrangements that SBI and SIDBI mobilized to reach out to SMEs, the approaches developed by these two lenders are very different:

- In spite of its extensive network of (now nearly 10,000) branches spread across the entire country, SBI relied exclusively on its Central Office (headquarters, located in Mumbai) for designing, managing, and staffing Project Uptech. As and when target clusters were identified in any state/region in the country, selected personnel from the Project Uptech central team were relocated to a branch in the vicinity of the target cluster while continuing to report directly to the Central Office in Mumbai. This approach was distinct from the standard SBI practice, wherein the regional offices and branches implemented policies made by the Central Office.[5]

- SIDBI, with less than 40 establishments across the country, relied to a large extent on about 900 other primary lending institutions (such as state-level financial entities, branches of other public sector banks, and cooperative banks) to increase its outreach to SSI units and SMEs.

Yet the two approaches had similarities (see figure CS5.1), as they clearly recognized the following:

- High transaction costs in relation to loan transaction size make it necessary to find solutions that reduce the cost per enterprise and/or per transaction.

Figure CS5.1. Cluster Lending Approach Adopted in India

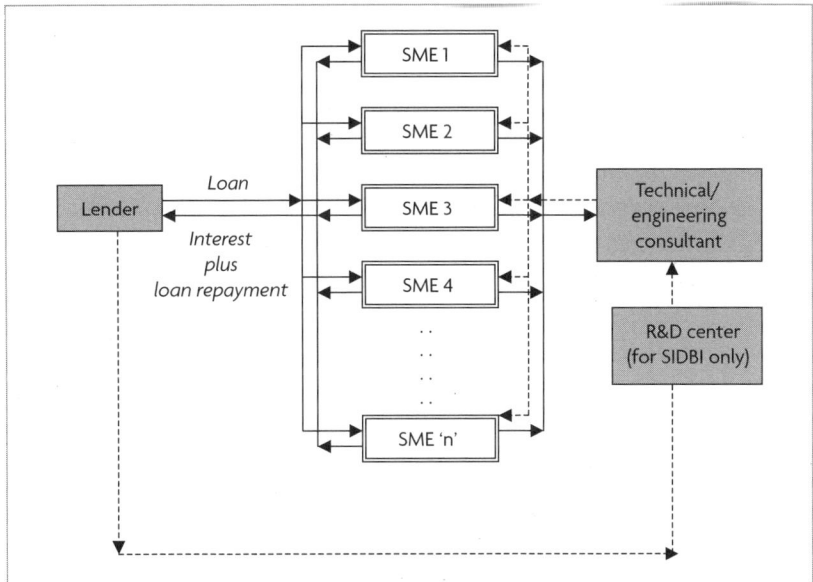

Source: Authors.

- Improvement in energy efficiency often is a direct consequence of broader technology upgradation and need not be the sole objective of lending.
- Technology solutions for SMEs are seldom (if ever) readily available off the shelf, even if a large number of SMEs need the same or similar solutions.

As a result, SMI and SIDBI took various measures to support some of their technology development work:

- SBI contracted external industry experts and (where necessary) research-demonstration-development organizations.
- SIDBI created a specialized agency, the Technology Bureau for Small Enterprises (TBSE).[6]

Project Example: Technical Experts for the Auto Cluster at Pune. In its simplest form, the lender selects technical experts in the target industry, and helps them to work closely with the SMEs (see figure CS5.1). The case of the auto component industry cluster in Pune (as well as the neighboring areas of Aurangabad, Nasik, and Ahmednagar) is a typical one. The process followed by SBI may be described as follows:

- The Project Uptech team catalogued the opinions of SME entrepreneurs in the cluster as well as their requirements, and identified possible organizations and experts to form an expert panel.
- With its expert panel in place, Project Uptech launched the implementation in the auto component industry cluster in January 1996.
- In close consultation with the entrepreneurs and SME workers, the technical experts identified a set of activities and investments required to enhance the competitiveness of the SME units. In so doing, the technical experts also helped the SME units to move from making simple components to subassemblies (a change that was being increasingly required by the newly emerging auto industry in India in the mid- to late 1990s).

- A total of Rs 96 million of loans were made by SBI to the 42 SME units over a four-year period (an average of over US$5,000 per SME unit).
- The borrowing SME units used the sanctioned loans to pay for equipment, machinery, or services as was found suitable for the particular SME. Typically, a SME unit used its loan to cover costs of one or more of the following:
  - studies conducted by technical experts
  - other studies (such as an energy audit in some cases)
  - IT equipment (for better management information systems)
  - specific equipment, such as a special-purpose shearing machine with automatic feed

Technical Assistance in Identifying Additional Technology Options. In cases where the baseline survey (by bank personnel and industry experts) indicated that a few specific technology interventions were required but were not readily available off the shelf, specialized R&D institutions could also be called into service. One such instance was the Firozabad Glass cluster, where the SMEs experienced high wastage (through glass breakage and rejections) as well as poor quality.

SBI's Project Uptech coordinated with various government agencies including the Centre for Development of Glass Industry and the Central Glass and Ceramic Research Institute. Several training programs and seminars were conducted as well as SME-level studies that covered more than 20 percent of the registered SMEs in the cluster. The focus was on appropriate technologies for process improvement, energy efficiency, pollution abatement, low-cost automation, introduction of management information systems, and above all, creation of an improved working environment.

Likewise, SIDBI appointed the National Productivity Council to work on reducing heat losses from locally fabricated furnaces in more than 1,000 aluminum, brass, and steel utensil manufacturing units in Jagadhari (Haryana). The cluster utilized wood and fuel in poorly designed and inefficient furnaces that had been fabricated by local masons, with the entire cluster having virtually the same furnace designs. Replacement furnaces designed by the National Productivity Council helped the companies to realize energy savings of upwards of 20 percent. The organization was thus similar to SBI's, except for the use of an R&D institute in addition to technical engineering consultants (see figure CS5.1).

Recent Energy Efficiency Cluster Lending Initiatives

The combined experience from SBI's Project Uptech and SIDBI helped provide a good base for the Indian banking community to launch energy efficiency–focused cluster lending programs. The trend was reinforced with the joint policy directive of the Government and the Reserve Bank of India in August 2005, which urged banks to increase credit to SMEs.

Prompted and encouraged by the Three Country Energy Efficiency Project, five banks have now formulated energy efficiency schemes targeted at SMEs.[7] In chronological order of the date of scheme sanction, the banks include:

- State Bank of India
- Canara Bank
- Union Bank of India
- Bank of India
- Bank of Baroda

KEY ISSUES IDENTIFIED FOR ENERGY EFFICIENCY LENDING TO SMEs

Key factors that contributed to the success of the cluster lending schemes have emerged from the initial implementation experience:

- **Active role played by bank Head Office.** The Head Offices of all five banks recognized the medium-term benefits of financing energy efficiency projects, and therefore provided relevant guidelines and other assistance to implement energy efficiency lending programs within their branch offices. This high-level support has proved critical in establishing any energy efficiency program at commercial banks.

- **Toward credible energy audits.** Recognizing further that credible energy audits are the basis for lending through the energy efficiency schemes, the banks have also tried to empanel selected energy auditors and ESCOs to ensure a high level of quality and reliability in technical assessment. For instance, SBI conducted a detailed review of the background and experience of energy auditors, ESCOs, and other energy consultants empanelled by the Bureau of Energy Efficiency (BEE), the Petroleum Conservation and Research Association, and others to identify a panel of their own. Nearly 30 energy auditors, ESCOs, and energy consultants were thus empanelled by SBI. Other banks have followed similar approaches. For example, the Union Bank of India's scheme states that the energy auditor should be approved by IREDA; and the Bank of India's scheme states the energy auditor should be approved by BEE or the Petroleum Conservation and Research Association or should have done similar work for SMEs. However, there is still a gap in understanding between what the banks consider suitable and the technical assessment undertaken by energy auditors. Further work is being undertaken to prepare standardized templates for energy audits to help bridge this gap.

- **Appraisal and loan acceptance.** Given the small loan sizes and the need to reduce transaction costs, appraisal processes have been simplified, while upholding the underlying bank needs for evidence of good credit profiles. The procedure generally is to formulate rules of thumb and guidelines for eligibility that personnel in branch offices can readily follow. For instance, SBI, Canara Bank, and Union Bank of India restrict the energy efficiency scheme to their existing SME clientele of a certain minimum credit score—which means that the borrower is well known to the lender. Combined with steps taken to ensure credible energy audits, the loan appraisal process is shortened considerably by this restriction.

- **Comfort from previous efforts.** The success of some of the early cluster lending programs of SBI and SIDBI certainly helps in extending the energy efficiency schemes. However, energy efficiency improvements were not the stated objective of the broader lending supported earlier. To that extent, the "demonstration effect" may be somewhat muted. In view of this reality, the intention of at least the SBI, Canara Bank, and Union Bank of India schemes is to offer some initial incentives for some projects by subsidizing "soft costs" (that is, costs of energy audits, preparation of detailed project reports, and so forth) to increase specific experience with energy efficiency investments.

LESSONS LEARNED AND THE WAY FORWARD

For the most part, each specific cluster project has been considered a success by SBI and SIDBI and perhaps the entire banking and development community. However, the entire concept has not taken off in the manner that was initially envisaged.

In the case of SBI, Project Uptech has remained a head-office-centered activity that has not become part of the mainstream lending business. In view of the recent emphasis on SMEs accorded by the Government, SBI is in the process of realigning its short-term priorities (of increasing business with SMEs and reducing bad debts from SMEs) with its longer-term objectives of encouraging SMEs to become modern, competitive, and financially sound.

On the other hand, although TBSE continues to increase its visibility and activity levels, it has found it difficult to spearhead cluster project developments. TBSE's main activities are as follows:

- information dissemination on new technologies and awareness enhancement initiatives
- facilitating matchmaking between technology suppliers and users
- providing support to SSI units on export of their products
- working with SIDBI to raise funds and syndicate loans for borrowers
- other support services such as arranging consultancy services where required, representing SSI business interests in international meetings, and so forth

Nevertheless, the benefits of the clustering approach have been demonstrated fairly clearly in a number of cases spread across the country and over a period of more than a decade. Key activities which can be undertaken for future scaling up of energy efficiency lending to SME clusters include:

- devising additional energy efficiency lending schemes
- making efforts to market the new and existing schemes
- empanelling certified, reputable energy auditors
- developing standardized energy audit templates
- developing in-house skills, and/or identifying a pool of external experts to assist on technical matters in the appraisal process

NOTES

1. SMEs in India are defined to comprise two broad categories: (i) small-scale enterprises whose total fixed assets in plant and machinery do not exceed Rs 30 million (~US$670,000); and (ii) medium-scale enterprises whose total fixed assets in plant and machinery range from Rs 30–100 million (~US$670,000–2.2 million).
2. In a now-classic article in regional science, Ann Markusen (1996) differentiates among typologies of industrial clusters. Each cluster type would present different challenges and opportunities for an energy efficiency lending strategy.
3. For instance, as of 2002/03, there were about 3.6 million SSI units with a workforce of nearly 20 million, which accounted for nearly 40 percent of industrial value added and about a third of total exports.
4. At present, the fiscal incentives to SSI units include (among others) the following: (i) complete exemption from—or lower rates of—excise duty payment obligations; and (ii) reduced company tax rates, for instance, in general allowing a deduction of 25 percent of profits in the first 10 years of operation for tax computation purposes. For SSI units located in designated backward areas, there is a further incentive in the form of elimination of tax payment obligations (i.e., tax holiday) for the first five years of operation. Other incentives have included product reservations, wherein certain products are allowed to be manufactured only by SSI units.

5. We are given to understand that SBI is moving toward a somewhat more centralized system, wherein (i) the Central Office in Mumbai would engage in policy making and monitoring and evaluation work; and (ii) the 14 local head offices (LHOs) would be responsible for the operations of various regional and local branches within their jurisdictions. Therefore, for most banking operations, (for example, extending term loans to industry), the policy framework, priorities, and so forth would be articulated at the Central Office, the decisions to sanction specific loans would be made at the LHOs, regional offices, and a few branches (where expertise resides). Most branches would merely act as a "post office"—that is, receiving applications and forwarding them to the nearest branch, regional office, or LHO for processing.
6. TBSE is a joint venture between SIDBI and the United Nations Asia Pacific Center for Transfer of Technology.
7. For more information on the various energy efficiency schemes, refer to the India Report of the Three Country Energy Efficiency Project, http://3countryee.org/reports.htm.

6. LITHUANIA ENERGY EFFICIENCY AND HOUSING PILOT PROJECT

INTRODUCTION

In former Soviet countries with cold climates, existing buildings are in dire need of structural and thermal renovations to preserve important assets, improve comfort, and generate substantial energy savings and reductions in heating bills.[1] Thermal renovations are costly, and the benefits of thermal renovations are only partially derived from financial savings. Thermal renovation of multifamily buildings poses special challenges, in that heat is usually not metered and cannot be controlled by the consumer, who typically pays for heating service on a fixed-fee basis. Organization of effective homeowners associations (HOAs) is difficult because there is no tradition of self-management of common building areas, which are instead maintained by notoriously inefficient, municipally owned companies. Finally, homeowners are generally not creditworthy, mortgages are not widely available, and the overall banking sector is underdeveloped and unwilling to lend for these types of improvements.

All of these facts conspire to make financing of thermal rehabilitation difficult at best. In many countries in the region, substantial government grants of 80+ percent were the only financing available for thermal efficiency projects in the 1990s; and these grants were limited in scope. The Lithuania Energy Efficiency Housing Pilot Project (1996–2001) was among the first projects financed by the World Bank[2] to take up the challenge and finance energy efficiency improvements in residential buildings through a credit line for homeowners and HOAs.

INSTITUTIONAL ARRANGEMENTS AND RESULTS

Figure CS6.1 shows the setup of the project. Its major focus was the improvement of energy efficiency in residential multifamily buildings managed by HOAs.[3] World Bank funds were onlent through a commercial bank to HOAs and individual homeowners.

Figure CS6.1. Lithuania Energy Efficiency Project Institutional Arrangements

```
                        World Bank
                    Loan  │  US$ 10M
                          ▼
                 Government of Lithuania
                   ╱              ╲
                  ▼                ▼
         Ministry of Finance    Ministry of Construction/
                                Urban Development (now
                                    Environment)
                                        │
                 Credit line  On-lending ▼
                                   HUDF (Housing and Urban
                        TA         Development Foundation)
          30% grant                   Project Coordination
                                        
                                   TA and procurement assistance
                        ▼               ▼
                       Bank          Municipalities
                                    (Vilnius, Kaunas)
                       Credit      Rehabilitation of schools
                    agreements  TA
         Interest/Principal
                                   Regional Advisory Centers
                                    (Danish Gov. TA support)
                        ▼
         Borrowers for investments  ◄--- Advice on technical, legal, financial, and managerial issues;
         in residential buildings:       support to organizing HOAs and promoting project benefits
         HOAs; individual home
         and apartment owners          Energy consultants
                                   (Dutch Gov. TA support for training)
                              Audits,
                          implementation
                           supervision
```

Source: Authors

The program included critical technical assistance from donors such as the Dutch and the Danish Governments. A support network of advisory centers in five major urban areas provided free advice on technical, legal, financial, and managerial issues to HOAs and individual homeowners. Private energy consultants were trained to support HOAs with preparation, for example, of energy audits, and supervision of renovation projects, carried out by private contractors. A comprehensive public information program was conducted. These undertakings improved Lithuanian public awareness on energy efficiency issues and enabled 193 HOAs and 25 owners of single-family houses to implement various packages of energy efficiency measures in their buildings.

Financial Intermediation and Subsidy Arrangements

- Only one bank participated in the project, acting as an agent without committing its own funds or responsibility for quality of appraisal. It received a fee of 1 percent of disbursements and 3 percent of collections. Initially the product was offered only in Vilnius; after merging with another bank, this product was offered at all local branches.

- Loans were made in local currency with an 11 percent fixed interest rate, which was initially below the implied market rate (there was no long-term lending available then), minimum 10 percent down payment, no mortgage requirements for HOAs, and a maximum loan maturity of 10 years.

- Since January 1999, the government has provided a matching grant, calculated as 30 percent of the loan principal, but not exceeding LTL 50 (US$12.5) per square meter of living area. In addition to the grant there was a partial VAT exemption for HOAs and individual homeowners.

- All HOAs with a valid registration and no outstanding arrears for utilities were eligible to apply. Loan proceeds could be used only for energy saving measures and urgent repairs ensuring improved energy efficiency.

- The loan repayment was shared between homeowners according to an agreement made in the HOA General Meeting, typically according to apartment size. Existing heat subsidies to low-income households could be used to pay for a portion of energy efficiency investment loans owed by those households.

Energy Efficiency Investments and Results

Table CS6.1 summarizes the project results:

- Initially most projects were limited to renovation of heating systems, particularly rehabilitation of district heating substations in

Table CS6.1. Lithuania Energy Efficiency Project Results (HOA Component Only)

	1996–1997	1998	1999	2000	2001	Total
Total amount of loans to HOAs, US$	74,300	206,000	1,161,000	3,375,500	2,401,200	7,218,000
Amount of grant provided, US$	n.a.	n.a.	442,300	905,500	695,200	2,043,000
No. of projects implemented by HOAs	5	18	49	111	46	229
No. of HOAs advised	87	113	312	113	101	726
No. of buildings audited	46	54	141	66	24	331
No. of investment proposals prepared	27	45	134	75	23	304
Average loan, US$	14,850	11,460	23,700	30,400	52,200	31,500

Source: See note 1.
n.a.: Not applicable.

building basements, which provided space heating and domestic hot water services. After introduction of the grant, window replacement and wall insulation also became more popular.

- In the total of 229 projects, 113 substations were rehabilitated in residential buildings, windows were changed and renovated in 144 buildings, and roofs of 41 buildings and walls of 26 buildings were weatherproofed.

- In 96 monitored projects, investments varied from less than US$ 250 per apartment to more than US$3,500, with an average close to US$1,000. Energy savings showed a very large spread, from significant additional consumption to more than 50 percent reduction, with an average value for a normal year of 17 percent (without adjustment for comfort change). In reality many homeowners preferred increased comfort to savings and raised indoor temperature. A rough estimation showed that without this temperature increase the average energy saving would have been 25 percent.

- Based on a survey, 56 percent of respondents had their heat bills decreased and 48 percent reported improvement in housing quality.

- The average payback time of investments (without taking into account increased comfort or other imputed benefits such as reduced

maintenance costs, extension of asset life, or increased property values) amounted to 17 years; taking into account the grant and the VAT exemption, it was 12 years.
- There were no defaults.

Follow-up
After the World Bank project closed in 2001, the government continued the program, including the capital subsidies (up to 30 percent) and support for low-income household participation (covering debt service payments). The extended program was financed out of budget resources and loan reflows through 2005. From 1996 to September 2005, a total of 1,200 HOAs participated; 799 investment projects were prepared by energy consultants; and 712 loans were issued by a commercial bank and a total of about US$22 million equivalent invested in thermal rehabilitation measures, of which only one quarter, US$5.3 million equivalent, was financed by the original World Bank loan.

This program has now been converted into an overall apartment building renovation program that uses commercial loans and guarantee/credit insurance mechanisms. Five banks were selected to participate in the program. The new Apartment Building Modernization Program started in late 2005. As of March 2006, the Program has generated 87 projects (of which eight are under implementation), totaling about US$ 9.6 million in investments and about US$2.4 million in government support.

The Advisory Centers created with Danish funding under the World Bank project, now called the Housing Advisory Agency, continue to play the same role for information dissemination and consultation to homeowners, including help with applying for subsidies. There are two categories of subsidy:

(i) **capital subsidy** of up to US$16 equivalent per square meter of useful area (size of subsidy depends on energy savings and ranges from 10 to 30 percent of the total cost of investment)

(ii) **low-income subsidy,** applied to
   a. defray costs of initial down payments (no more than 10 percent) for the investment

b. credit insurance premiums

c. a certain part of debt service burden

The government also continued to introduce energy efficiency measures in its public building retrofitting programs. It borrowed from the Nordic Investment Bank for energy efficiency renovations of higher education, hospital, and other public facilities, and from the World Bank for school renovation in two projects.

LESSONS LEARNED

- Thermal retrofit of buildings is among the more difficult energy efficiency investment measures to implement. Considering only the benefits of energy savings leads to relatively long payback times and tends to make such investments difficult to finance on purely commercial terms. Additional benefits occur to the home owner, such as improved comfort and increased value of the dwelling, but these might not be easily turned into cash flows. From an economic/country point of view, thermal renovation has additional aspects and benefits that justify providing incentives such as capital grants, including improvement of communal spaces in multifamily residential buildings and of their physical integrity in general, increases in the economic lifetime of building assets, improved comfort, and reduced environmental impact of heating through reduced overall demand. Thermal energy efficiency measures should be considered in conjunction with other building renovation measures such as wall or roof renovation undertaken for different reasons. The additional cost of adding more or better insulation is fairly small in these renovation projects. Also, in some cases HOAs decided to add an additional floor to the building, and the income from selling or renting the additional space provided cash for loan repayment.

- Loans to HOAs are particularly challenging since they require a lengthy process of consensus building. Although HOAs didn't have to take a mortgage, loans to owners of individual homes required such collateral and uptake was therefore relatively small, with only 25 loans. The HOA loans provided a successful bundling mechanism, leading to a much higher average loan size and lower transactions costs per dollar lent. As a result of their involvement, HOAs have been strengthened and communal space is much better cared for than in many other countries in the region.

- Physical savings were smaller than expected (but, in many cases, increases in heat consumption were offset by domestic hot water savings that were much higher than expected); benefits materialized partially in terms of increased comfort. As a result, payback times tend to be on the long side. Underheating could thus become a problem for commercial investment in energy efficiency and deter investment by third parties such as ESCOs.

- The project initially had a limited impact on the expansion of commercial lending to the housing/residential sector, due mainly to the underdeveloped market for such borrowing. The design of the project did not meet the strategic interests of commercial banks, which were recovering from a major banking crisis in 1995–96. Although initially several banks had been targeted for participation, only one bank could be convinced to join the project through the concerted efforts of the government and World Bank team. With the continuation of government support in a follow-up program, lending for building renovation is now becoming a standard banking product.

- Considerable technical assistance is required to provide technical, administrative, and marketing services to ensure project uptake and successful implementation. The advisory centers and energy consultants, initially financed by donor contributions and providing free services to potential borrowers, are continuing to provide these services, but are currently doing so on a fee-basis. Advisory centers and energy efficiency consultants were key in providing project marketing, technical assessment, appraisal due diligence, and training for HOAs.

- A strong project coordinator, the Housing and Urban Development Foundation (HUDF), played a central role in local management of the project by assisting and supporting all key players in the project with substantial donor grant funds and foreign consultants. HUDF helped publicize and promote the project by supporting the establishment and organization of HOA support through regional advisory centers with help from energy efficiency consultants.

Some of the other MDB-financed projects supporting commercial lending for energy efficiency investments, particularly in residential buildings, have incorporated some of these lessons. The new EBRD-financed credit line for residential energy efficiency in Bulgaria is targeted at individual home and flat owners who are taking out personal loans from one of four participating banks. Eligible technologies and suppliers have been defined. The loan program is accompanied by TA for technical and marketing services and a 20 percent grant, both financed from the Kozloduy International Decommissioning Support Fund. As of mid-2007, more than 8,600 loans have been provided for a total surpassing US$16 million.[4] In Hungary, the IFC-supported HEECP has resulted in new banking products for lending for gas heating and blockhouse renovation (see Case Study 2). Both types of investment benefit from government grants for energy efficiency measures, which have a long tradition in Hungary.

NOTES

1. Unless otherwise noted, information in this case study is based on World Bank (2002) and HUDF (2002).
2. This project was preceded by the 1996 Russia Enterprise Housing Divestiture Project where World Bank funds were onlent by the government to several municipalities for thermal renovation of municipality-owned residential buildings.
3. The project also included financing for energy efficiency renovation of schools in several municipalities for a total of US$4.7 million and TA for institutional development and strengthening.
4. Exchange rate: 1 US Dollar = 1.45349 Bulgarian Lev (as of June 29, 2007). See http://www.reecl.org.

7. CHINA'S FULL-SERVICE ESCOs

INTRODUCTION

China's ESCOs, better known within China as Energy Management Companies (EMCs), are a rapidly growing industry. With the energy performance contracting investments of Chinese ESCOs estimated at about US$280 million in 2006, involving over 60 different ESCOs, China's ESCO industry is clearly one of the most successful adaptations of the energy performance contracting mechanism in the developing world. Profitability of the mechanism in the Chinese market has been clearly demonstrated, and is the key reason for continuing sharp growth. The full-service, shared savings model that is common in China offers a compelling package to customers. A drawback of this mechanism, however, is the need for large amounts of loan financing to the ESCO themselves, in order for them to maintain growth.

THE CHINESE ESCO MECHANISM

As shown in figure CS7.1, the full-service, shared savings ESCO model as practiced in China places the ESCO in charge of all aspects of project development, financing, and implementation. Described as "service from the head to the tail of the dragon," the ESCO manages the following functions: (i) completes a diagnostic analysis and project identification; (ii) reaches agreement on the expected energy cost savings to be generated by the project with the client, and a payment stream accruing to the ESCO from those savings by the client; (iii) finances all or most of the project investment, but usually with some type of repayment guarantees arranged by the client; (iv) completes project procurement and equipment installation; (v) undertakes commissioning, testing, training, and maintenance according to prior agreement; and (vi) receives payment according to the payment stream agreed up front, unless the project does not perform as expected.

Figure CS7.1. China's Full-Service Shared Savings ESCO Model

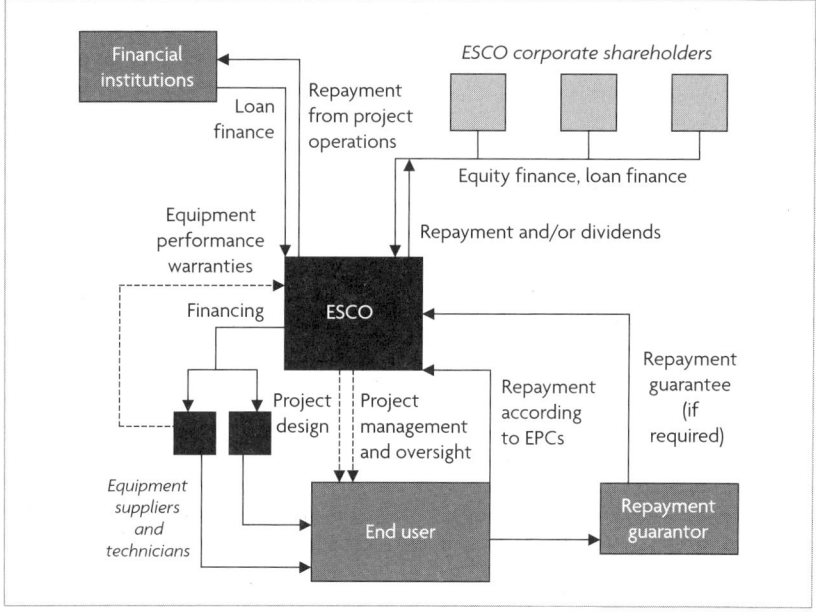

Source: Authors.

China's ESCOs—defined as any company that implements energy performance contracting projects—totaled more than 60 at the end of 2006. Most are shareholding companies.

The two main markets where ESCO businesses are active in China today are industry and commercial buildings. Industrial renovation projects are generally equipment-focused, including, in particular, boiler renovations, technology upgrading in combustion systems, renovation of kilns and furnaces, waste heat or gas recovery and use, motor drive system renovations, cooling system replacements, internal power supply renovation, and introduction of automatic controls. Commercial building projects focus primarily on heating, ventilation, and air conditioning (HVAC) system renovations and innovations, although lighting system renovation projects also exist. Projects focus on fuel savings—especially direct savings in coal use—as well as electricity savings.

Most, but not all, Chinese ESCOs are quite specialized in the technical products they offer. Many are, or are related to, equipment vendors, and quite a few have specific patented technologies that are a central business focus. Even the larger and broader companies tend to have technical specialties. In marketing their energy performance contracting businesses, therefore, most ESCOs will not complete comprehensive diagnostic work, such as energy audits of client facilities, but rather will target their marketing to companies that are likely to be interested in the specific technical products of the ESCO, and are in a position to make repayment. Some of the more mature ESCOs, however, are now beginning to offer more comprehensive diagnostic services to customers who are interested.

Typical shared savings EPCs in China are simple, especially compared with North American contracts (see figure CS7.1). Project content and the responsibilities of the parties are clearly defined. The energy cost savings expected from the project are estimated and agreed, using prevailing energy prices. The ESCO finances the project, and is paid from a large share of the initial energy cost savings estimated. Payment streams are clearly defined in the contract, based on the percentage share of the ESCO in the estimated savings (around 80 percent is common). Contracts may define energy savings testing procedures, but often one test at equipment commissioning stage is sufficient. Payment streams to the ESCO are almost never revised: as long as the equipment functions properly, payments are based on the energy savings stipulated in the contract. Contract tenure is relatively short, very rarely more than 1–3 years (except for contracts incorporating long term maintenance by the ESCOs). There is interest among some Chinese ESCOs in more sophisticated energy savings monitoring and verification methods, especially for possible use with customers who have little technical understanding of the projects involved. However, stipulated savings and fixed repayment schedules are the mainstay of the Chinese ESCO industry today.

As the Chinese ESCO industry develops, different variations of energy performance contracting are emerging. Guaranteed energy savings contracts, especially for building energy efficiency projects, are now common. In these contracts the ESCO guarantees project energy savings performance, but financing is arranged by the client. Chauffage arrangements (see Case Study 11), or contractual arrangements where

ESCOs assume energy bill payment responsibilities, are also emerging. The industry and market are highly dynamic, yielding diversity and flexibility in approaches.

As businesses that provide the financing for projects, probably the single most important function of the full-service ESCOs is ensuring repayment from clients. Repayment considerations heavily affect project scope (for example, there is a heavy focus on projects where breakeven repayment is achieved as quickly as possible) and client selection (for example, focus on client creditworthiness and/or clients where trust-inspiring relationships may already exist). The three large pilot EMCs created in 1997 usually require clients to arrange repayment guarantees from a third party (such as a guarantee company, a large client shareholder, or local government) as part of the EPC package. These EMCs also have formal internal mechanisms for reviewing the repayment security of each project. In Shandong EMC, for example, all projects developed must be sent to a separate financial review and appraisal department for formal approval before contracts can be signed. Over eight years, only about a half dozen of some 330 projects implemented by the three companies have suffered from major defaults, but this success has been achieved only through active, careful management.

Arranging financing for project implementation is also a core task for full-service ESCOs. The World Bank played a central role in provision of finance for the three pilot ESCOs. However, as shown in table CS7.1, indirect support through the ESCO Loan Guarantee Program for commercial bank loans (Case Study 1) and independently raised financing are now playing increasingly large roles.

PROJECT SUPPORT FOR ESCO DEVELOPMENT IN CHINA

China's ESCO industry was developed basically from scratch through strategic efforts of China's Government, with support from the World Bank. Introduced in 1994–95 as a potential investment concept, energy performance contracting caught the imagination of key energy efficiency policy makers as a possible mechanism to promote energy

Table CS7.1. China ESCO EPC Project Investment 2005/2006 (Millions of US$)[a]

	2005	2006
EMCA members		
Three pilot EMCs [b]	33.4	30.3
Projects guaranteed by I&G	26.8	18.6
Other EMCA member projects	137.1	179[c]
Total	197.3	227.9
Non-EMCA members	48.4	52.0
Total	245.7	279.9

Source: World Bank estimates.
a. Exchange rate: Y 8/US$1 for 2005, Y 7.8/US$1 for 2006.
b. Beijing, Liaoning, and Shandong EMCs, supported under the World Bank/GEF China Energy Conservation Project.
c. US$153.4 million from data reported from EMCA members, excluding the three pilot EMCs, plus US$25.6 million extrapolated from random sample survey results.

efficiency investment commercially through the growing market economy at a time when China's economy was transitioning from a planned to a market-based system. There was no known experience in energy performance contracting in the country at that time. During 1996 and 1997, the World Bank and the government prepared the World Bank/GEF China Energy Conservation Project, approved in 1998, to introduce the energy performance contracting investment mechanism to China.

The China Energy Conservation Project provided support for the establishment of three new pilot ESCOs, and their efforts to try to adapt and develop energy performance contracting in the Chinese market. The project included (i) US$5 million of GEF grant support to each company for development of initial projects, and (ii) US$21 million of IBRD loan funds, onlent to the companies at commercial rates with the ESCOs covering all foreign exchange and repayment risks, for scaling up their business. Shareholding companies were specially formed for the project in Liaoning and Shandong Provinces and Beijing Municipality, with shareholders consisting mainly of other publicly owned companies. A major TA program for the new ESCOs and funding for a few initial pilot projects was financed by the European Commission, beginning in June 1997, which gave the program a very important jump start.

This project has succeeded in meeting its goals. The three new ESCOs have adapted energy performance contracting profitably to the Chinese market, creating strong operational examples that are now being replicated by many. By the end of 2006, the three ESCOs had implemented about 350 energy performance contracting projects, with investments totaling about US$170 million. Financial performance and profitability of all three companies was strong.

The majority of projects implemented by the three pilot ESCOs are small-sized, with more than two-thirds below the average project size of about US$350,000. As shown in figure CS7.2, many different types of energy efficiency projects have been implemented by the three EMCs between 1997 and 2006. Building renovation and boiler renovation/ cogeneration are the most common in terms of investment volume, both with about one quarter share. It is worth mentioning that the three EMCs concentrate on different sectors. Almost three quarters of the investments of the Beijing EMC are in building renovation (mostly HVAC), while the EMCs in Liaoning and Shandong invest much more heavily in the industrial sector with more than 60% of investments in boiler/cogeneration, kiln/furnace and waste heat/gas recovery.

Eying the commercial success of the first three ESCOs, additional ESCOs began to develop in China around 2000. Technical assistance from the U.K.'s Department for International Development funded a major national effort to provide emerging knowledge on the business to new companies, train professionals, develop initial financing concepts, and establish China's new national EMC (ESCO) Association (EMCA). The Second Energy Conservation Project, with US$26 million of GEF financing through the World Bank, was approved in October 2002 with the objective of scaling up China's ESCO industry nationwide. This project includes (i) a major EMC loan guarantee program, backstopped with GEF funds (see Case Study 1); and (ii) training, technical assistance, and policy development support for the emerging ESCO industry through EMCA. Details about the implementation of the guarantee program are provided in Case Study 1. EMCA has more than 100 members, to which it provides a wide range of support.

Figure CS7.2. Types of Projects Implemented 1997–2006 by Three Chinese ESCOs

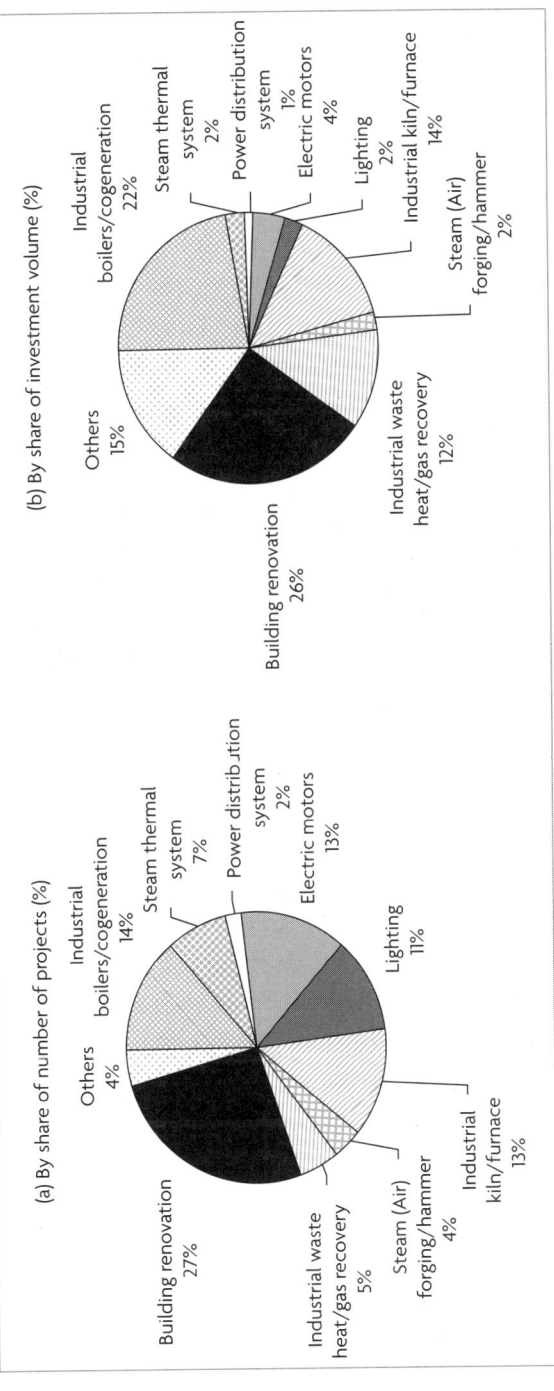

Source: NDRC/WB/GEF/Project Management Office (2006) Energy Conservation Projects: Case Studies of Chinese ESCOs. China Economic Publishing House, Beijing.

ADVANTAGES AND DISADVANTAGES OF THE CHINESE FULL-SERVICE ESCO MODEL

A key advantage to this investment mechanism is its attractiveness to end users. The ESCO undertakes the main work, provides off-balance-sheet financing, and shoulders the technical and financial repayment risk. The mechanism also has been proven to generate strong financial returns to both ESCOs and clients, as the financial viability of the typical projects is very strong, providing sufficient returns for both entities to share. The Chinese business community has proved adept at adapting the basic principles of energy performance contracting to different circumstances and markets, with a wider variety of variations. The potential for ESCO industry growth and long-term sustainability in China is considered quite high.

A major disadvantage of the full-service, shared savings approach is the need for large amounts of capital to maintain ESCO growth. Raising capital for investment is difficult for the small companies that dominate the ESCO industry, and these companies can quickly reach loan and equity financing ceilings, hampering their ability to grow strongly while they await repayment proceeds. The Guarantee Program is not a panacea for this problem. The guarantee company facilitates the business by acting as a financial intermediary that understands both the ESCOs and banks well, and by introducing innovations catering to the business. However, the guarantee company must observe regular commercial norms, and cannot shoulder undue risks of lending to overleveraged small companies. Means to overcome this problem include the following: (i) increasing use of contracting variations where mobilization of financing is shared more with clients; (ii) development of larger, financially stronger ESCOs (a number of major corporations are now beginning to enter the market); and (iii) further innovation in development of loan structures for the business. One such innovation could be increasing linkage of loan and EPC agreements to take advantage of end-user creditworthiness. Another innovation could be mechanisms to ring-fence project repayment streams (for example, by recognition of ESCO accounts receivable as loan collateral, or development of bank recourse to repayment streams in escrow accounts). In any event, it also should be recognized that the full-service ESCO industry in China still continues to grow sharply despite this problem. Where profitability exists, Chinese companies have been finding both formal and informal ways to raise capital.

Another disadvantage of the full-service, shared savings ESCO model is its complexity and the need for a blend of diverse skills for successful operation. Required corporate skills include strong technical skills (which underline the ESCOs' value added to many customers), as well as strong financial management and business skills. The key to financial success is ensuring repayment from customers, and hence the ability to shrewdly assess financial risks is paramount. This model cannot be easily duplicated in a completely new environment. It is not easy for the business, financial, and regulatory communities to quickly understand, accept, and trust. The Chinese full-service ESCO is partly a technical design enterprise, partly an equipment procurement and installation service, and partly a financing institution. In China's case an inability to easily categorize the business caused initial confusion and great argument among taxation authorities, financial auditors, legal advisors, and financial system regulators.

LESSONS LEARNED FROM THE CHINA ESCO PROMOTION PROGRAM

Three of the many lessons learned from this program are summarized briefly below.

Strong, consistent public sector support was essential for the ESCO industry to get started and take hold in China. The Chinese Government undertook ESCO development as a type of strategic task from the beginning. Strong government support was critical for the design and implementation of the various programs to spur ESCO development. This included support for development of a World Bank loan proposition that was initially quite risky. In the early years of ESCO operations, senior-level government support was critical to overcome a range of serious issues concerning taxation, auditing, and ESCO legal status, including various claims that the business was "illegal." Increasing policy support, raising public awareness, and support for the development of the new institutional mechanisms of the Second Project also have been critical for scale-up.

The entire effort could not have been launched without the substantial support of GEF grant financing, especially for the pilot ESCOs. The demonstration that energy performance contracting could work profitably under Chinese conditions implemented by Chinese companies provided a solid foundation for all future ESCO development in the country. The equity investors for the three pilot companies were very clear that they would not make any investment in such an untried, unconventional, and foreign scheme without significant matching concessional financing. The Chinese Government was also clear that it was unwilling to provide counter-guarantees for a World Bank loan for such a risky project without substantial GEF support. Even so, the majority shareholder of one of the three ESCOs balked at the last minute due to the riskiness of counter-guarantees required of it for the World Bank loan for what the shareholder (correctly) described as a business with no track record.

In this case, the decision to develop the industry in stages has proved successful so far. In the first stage, provision of a line of credit through the public sector removed the financing constraint from the pilot ESCOs, allowing them to focus their efforts on making the performance contracting mechanism work. As ESCOs now seek to overcome financing issues in the second stage, then, the basic operating mechanism has already been proven.

Flexible adaptation to local conditions has been critical. Chinese practitioners learned the basic concepts of energy performance contracting from abroad, and listened closely to experiences and advice from the ESCO community overseas. However, business practices, contracts, corporate structures, project scope, risk management tools, and marketing methods all have developed distinctly Chinese characteristics—at times running counter to advice initially given from foreign experts. This adaptation continues as the industry continues to evolve.

Development of an ESCO Association has been helpful. EMCA serves as a critical forum for ESCOs in China to exchange knowledge and provide mutually reinforcing support to each other, a vehicle for ESCO industry representation at home and abroad, and an institutional mechanism for publicly financed training support for new entrants to the business.

8. ESCO DEVELOPMENT IN THE UNITED STATES AND CANADA

INTRODUCTION

The ESCO concept, which was developed in North America in the late 1970s, is often presented as a model delivery mechanism for energy efficiency retrofits in developing countries and emerging market economies. Thus, understanding the concept, the drivers, and the evolution of the North American ESCO market can be very useful in evaluating whether and how the generalized ESCO model might be adapted into a viable energy efficiency retrofit delivery system in markets with vastly different institutional and organizational environments.

This case study addresses primarily the development of the U.S. and Canadian ESCO industries, which originated and still have their main focus in the public buildings sector. The case study also presents information on how state and province-level utility DSM programs helped the U.S. and Canadian ESCO industries develop both public and private sector applications.

The Importance of Public Building Energy Efficiency Programs for ESCO Development

In both the United States and Canada, the federal, state/provincial, and local governments utilized ESCOs as a way to overcome barriers preventing implementation of energy efficiency projects in public sector buildings. These projects aimed to reduce energy expenditures and meet public policy goals of reducing energy use. Governments needed firms with energy efficiency expertise and project implementation experience to develop projects that would not require an increase in capital or operating budgets, but instead have neutral cash flow consequences while delivering cost-effective investments. Energy performance contracting (EPC) was developed as a concept to deliver those objectives.

The success of ESCOs and EPC in public buildings in North America in the 1970s and 1980s gave those energy efficiency delivery mechanisms credibility vis-à-vis government officials and financial institutions in other applications as well. The development of the ESCO industry in both the United States and Canada was due, in a significant part, to the encouragement, facilitation, and partnerships of government at all levels. EPC by ESCOs has become a major mechanism for federal, state/provincial, and local agencies to reduce energy use and to save operating costs.

U.S. ESCO DEVELOPMENT

The early U.S. ESCO industry received a major initial push from the Federal Institutional Conservation Program created in the late 1970s. This program targeted energy efficiency improvements in state and local schools and hospitals and was administered by state energy offices. Federal grants were provided for both technical assistance and capital investment, though the capital grants also required an equal share from nonfederal sources. Importantly, the federal rules allowed the nonfederal share to include payment from performance contracting with private partners. This led state and local officials to seek out such partners as a means to provide the nonfederal capital and the expertise for developing and implementing the energy efficiency projects.

Public schools and hospitals provided excellent energy efficiency project sites. They were both numerous and similar to other schools and hospitals around the state and around the country, thus providing substantial potential for wide-scale replication by ESCOs over a large potential market. The innovations and the lessons learned from early program implementation in the United States benefited significantly from the program being administered by 50 state energy offices that were driven by a "use or lose" federal funds perspective. Implementation across so many states increased the scope for diversity in finding innovative solutions. These energy offices were and remain advocates of EPC by ESCOs in public buildings.[1]

The developing ESCO industry and the public officials wanting to implement energy efficiency investments were initially hampered by the nonavailability of public funding, by restrictions posed by government budgeting rules, and by equally restrictive government procurement rules. To complicate matters further, some government officials, especially those with fiscal responsibilities, did not understand the concept that "the capital for the energy efficiency measures is in the savings from future operating costs." Fortunately, ESCOs and the EPC concept had advocates in state government willing to work to change the government policies and rules to encourage performance contracting through ESCOs. Thus, by the time of this writing, most U.S. states (over 40) have enacted comprehensive legislation promoting EPC in schools and government facilities.

A good example of these developments is Ohio, which in 1985 enacted legislation allowing school districts to purchase energy conservation measures on a multiyear installment basis and to increase the portion of a district's net indebtedness that could be used for energy conservation measures. Such indebtedness now can be incurred without a vote of the people, if the funds are used for energy conservation measures that self-fund over a 10-year period. The new legislation jump-started school-ESCO projects; and in the first five years ESCO projects worth more than US$131 million were developed in 167 school districts in Ohio.

The U.S. federal government passed additional legislation from the mid-1980s through the 1990s facilitating the use of ESCOs to provide energy efficiency retrofits in federal facilities. The Comprehensive Omnibus Budget Reconciliation Act of 1985 encouraged federal agencies to implement energy efficiency projects through shared energy savings contracts. Subsequent legislation provided incentives to federal agencies to participate by letting them retain and then use a portion of the foregone energy costs to expand ongoing programs.

The early EPC projects in U.S. agencies were based upon sharing the energy cost savings between the ESCO and the customer, with the ESCO undertaking the borrowing. (This is the shared savings model discussed in Part I; see figure CS8.1 for an outline of the deal structuring involved.) This concept was largely abandoned in the

1980s due to excessive litigation over claimed savings and was eventually replaced by having the government client take out the loan for financing the energy efficiency investment, backed by the ESCO's guarantee of the related energy savings. (This is the guaranteed savings model, also discussed in Part I; see figure CS8.2 regarding that

Figure CS8.1. Shared Savings Contracting Model

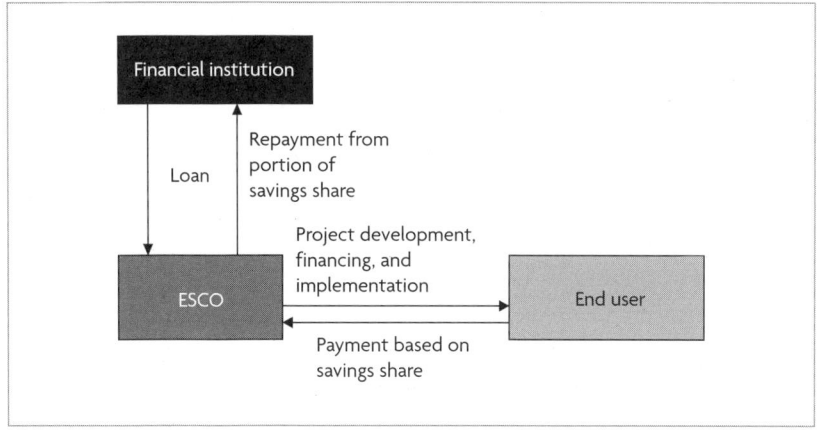

Source: Authors.

Figure CS8.2. Guaranteed Savings Contracting Model

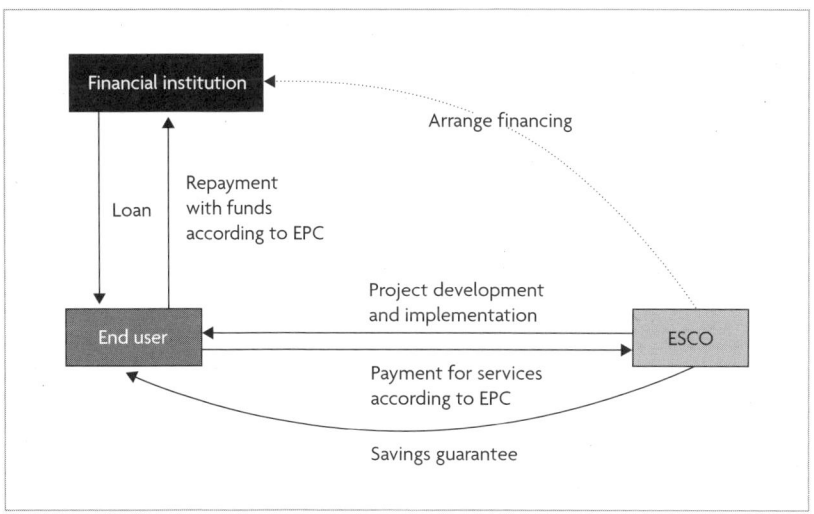

Source: Authors.

kind of deal structuring.) This latter model continued to evolve and to develop an increasing number of financing options.

Later on, guaranteed savings ESCO projects financed through tax-exempt leasing arrangements would become an important financing mechanism for energy efficiency investments in the institutional sector[2] overall in the United States. The advantages of this particular financing mechanism are as follows: (i) the projects can be included in the operating budget rather than in the capital budget; (ii) they require no voter or legislative approval; (iii) the projects can be established as a master lease, allowing continuous drawdown of funds as needed; (iv) the projects are long term (10–15 years); and (v) they can be more easily approved within government procurement systems than can a capital budget item.

Complex U.S. federal procurement rules and processes hindered federal agency use of ESCOs initially; and successful project development in the early years required an active partnership between federal agency staff and ESCO staff. As more projects were developed, a clearer path through the procurement maze of each agency began to be understood and to become "codified" into standard practice by a widening range and number of agencies.

Agency and agency staff incentives were extremely important (as highlighted in Chapter 4 of Part I) in gaining agency staff's initial interest and in continuing their focus on developing projects with ESCOs. For example, when Department of Defense base commanders figured out that they could use some of their share of the energy savings on welfare, morale building, or recreation activities, this presented a powerful incentive for commanders and staff to work with the ESCOs in identifying and developing energy efficiency projects.

Currently, the Federal Energy Management Program (FEMP), operated by the Department of Energy, is responsible for coordinating federal agencies' energy efficiency activities. FEMP promotes use of ESCOs and EPC by facilitating the navigation of procurement processes and maintaining lists of approved ESCOs. FEMP also developed the Super Energy Savings Performance Contracts (Super ESPCs) for use by federal agencies to implement comprehensive energy saving projects. This concept allows federal agencies to bypass procurement procedures and deal directly with a prequalified ESCO on energy projects.

A U.S. federal agency can have many projects with a particular ESCO under an umbrella contract. FEMP selects a limited number of ESCOs to be eligible for Super ESPCs through a competitive process. Then a selected ESCO conducts a comprehensive energy audit, identifies "bankable" measures to solve energy efficiency problems, arranges financing, guarantees a level of annual cost savings to the agency, and takes on responsibility for implementing the agreed measures. The agency pays for the measures utilizing a portion of the guaranteed savings. As of mid-2007, the Super ESPC concept had resulted in energy efficiency investments of US$1.9 billion by 19 agencies in 46 states.[3] Meanwhile, the Energy Policy Act of 2005 has extended the authority for super ESPCs until 2016.

The developments discussed above resulted in ESCOs becoming a formidable force in energy efficiency financing and implementation in the United States. According to a major survey and analysis by the Lawrence Berkeley National Laboratory,[4] the U.S. ESCO industry accounted for energy efficiency projects (institutional and private sector combined) of at least US$16–20 billion between 1990 and 2000 (underreporting of private sector projects likely makes the total higher). ESCO revenues from energy efficiency investments were estimated at about US$2 billion for 2000. The institutional sector accounted for about 75 percent of project investment, and private sector projects accounted for the remainder.

CANADIAN ESCO DEVELOPMENT

Through the Ontario Hydro Demand-side Management activities in the late 1980s and early 1990s, the Ontario government played an important role in stimulating ESCO development in Canada. As in the U.S. case, there was a particular focus upon institutional buildings.

Ontario Hydro's Guaranteed Energy Performance Program required utility customers to form partnerships with accredited ESCOs having proven track records in comprehensive energy efficiency retrofits. The targeted customers included government buildings, schools, hospitals, private retail and office buildings, and public and private multifamily buildings. Public buildings became a major focus of this program.

The ESCO industry in Canada received another very significant push starting in 1991 with the federal government's launch of the Federal Buildings Initiative (FBI). The FBI allowed federal departments to contract with ESCOs to develop and implement energy efficiency retrofits through EPC in federal buildings. Natural Resources Canada manages the FBI program by strongly promoting the concept to individual agencies, organizing training for agency staff on developing projects with ESCOs, providing model contracts and bid packages, and maintaining a qualified bidders list of ESCOs. Departments must submit their first contract exceeding Can$1 million to the Treasury Board for review. Once this contract is approved the department can contract without Treasury Board review for projects involving less than Can$25 million.[5]

FBI projects can utilize three types of EPC: (i) first-out, (ii) shared savings, or (iii) a combination of the two. Under first-out EPC, the ESCO retains 100 percent of the energy savings until the project is paid for or until the end of the contract, whichever occurs first. The ESCO declares its total investment costs and its markup toward profit and interest costs, with the markup considered part of the ESCO's cost basis. The customer retains all of the energy savings after the ESCO's costs are recouped.

Under the shared savings EPC, on the other hand, the department and the ESCO each receive an agreed percentage of the energy cost savings over the life of the project. The FBI allows the first-out and shared savings options to be used for different department facilities under a single contract. In practice, most departments opt for the first-out contract approach.

Under the FBI, 7,500 federal buildings and other facilities (through 2006) have been retrofitted, resulting in Can$240 million of private sector investment and annual savings of Can$33 million.[6] Although this program has evolved on a steady course for nearly 15 years, the ESCO industry in general is implementing significantly fewer projects now than at its peak in the mid-1990s, when it was performing roughly US$300 million worth of retrofits annually, mostly in the institutional buildings sector. There are two major reasons for the decline in activity: (i) many of the large and willing institutional facilities already have been retrofitted through previous projects, though some second-generation projects have been performed or are arising anew at those same sites; and (ii) the programs supported by utility DSM have been significantly reduced since the mid-1990s.

OBSERVATIONS AND LESSONS LEARNED

Government commitment was essential. The development and growth of the North American ESCO industry, reaching an annual investment level of US$2.5–3 billion by 2005, was due to government involvement at the federal, state/provincial, and local levels in both countries. Canadian and U.S. government policies and programs not only promoted and facilitated ESCO projects; they provided the ESCOs with attractive initial projects developed with creditworthy customers.

Numerous creditworthy public facility customers with replicable projects provided a great market. Government facilities, especially schools, hospitals, and offices, provided the ESCOs with projects having seemingly unlimited replication potential. Moreover, financing sound projects was not a significant issue—once the initial problems had been worked out—due to the highly creditworthy nature of the government customers. The newly discovered potential market of large numbers of creditworthy institutional customers with similar facilities and high energy saving potential allowed the North American ESCO industry to develop and grow. Thus, as procurement, budgeting, and public agents' incentives issues were resolved, availability of financing was rarely a deal breaker.

Public sector contracting and procurement are barriers that can be overcome by a government committed to energy performance contracting. Energy performance contracting by ESCOs in U.S. and

Canadian public buildings had to address and successfully navigate the detailed public sector contracting requirements and extensive public sector procurement processes. Public sector contracting and procurement can pose difficult barriers to new concepts such as energy performance contracting. Rules directed at controlling potential misuse of public funds dominate public budgeting and public procurement practices, to the severe detriment of productive efficiency in the public sector. In the public sector, misuse of public funds carries a much higher penalty for a public official than does missing an opportunity to reduce expenditures or to increase operating efficiency. Thus, the contract culture in the public sector differs significantly from the private sector, where seizing profitable opportunities constitutes a much greater force affecting procurement decisions.

The public sector contracting and procurement barrier had to be overcome in the United States on a state-by-state and agency-by-agency basis. Legislation and written rules for implementing the legislation differ from one level of government to the next, as do established procedures at the agency level. Model contracts, model procurement procedures, and model monitoring and verification requirements were needed to develop a smooth path through contracting and procurement systems in each case. Once one state had a functioning system in place for EPCs via ESCOs, other states were in a better position to adapt those innovations to their own circumstances.

This discussion also highlights the differences in the way the Canadian and U.S. EPC-ESCO interfaces developed, providing another stark example of time-path dependence. The Canadian public building ESCO development tended to operate top down, from the federal level down to the provinces, largely deriving from the implementation of the FBI program. The Canadian Treasury Board provided the energy performance contracting authority to the federal agencies, and Natural Resources Canada provided the model contracts and bidding package and training to develop contracts in their buildings. The United States, on the other hand, developed a bottom-up approach that grew out of the way that the state and local programs emerged.

The U.S. and Canadian ESCOs developed a niche market—public buildings. The U.S. and Canadian ESCO models primarily evolved from success in the public buildings markets. The characteristics of the public building markets (as discussed at selected points above) are as follows:

- high potential levels of cost-effective energy efficiency investments
- highly creditworthy customers
- customers wishing to avoid adding additional debt
- easily arranged financing
- customers who readily outsource many tasks
- customers with limited expertise in energy efficiency retrofits
- easy replication of basic projects among many similar customers
- active marketing and facilitation of ESCO projects by government officials
- partnerships with governments to develop the market by reducing identifiable barriers and by providing incentives to customers

The North American ESCO experience provides valuable contrasts—as well as complementarities—for energy efficiency financing experiences from China, India, Brazil, and other countries. The comparison is valuable because it comes from two highly developed market economies, which offers substantive contrasts in several respects with some of the comparator countries. However, the foregoing case demonstrates that, even in these two advanced market economies, much work had to be done to redesign incentives and rules of behavior by important actors in order to structure deals that would result in energy efficiency gains. Moreover, even with the similarities between their economies and with the powerful market connections between them, Canada and the United States found different pathways in building the different forms of ESCO industries that emerged.

NOTES

1. To help public officials interested in pursuing EPC and various financing options, the National Association of State Energy Officials (NASEO) provides a public building manual with links to many state programs. See http://www.naseo.org/committees/buildings/documents/NASEO_Public_Buildings_Manual.pdf.
2. As used in this case study, the term "institutional sector" includes government buildings, particularly educational and health care facilities, and public housing.
3. http://www1.eere.energy.gov/femp/financing/printable_versions/superespcs.html.
4. See http://eetd.lbl.gov/EA/EMP/reports/naesco_intl.pdf.
5. See http://www.oee.nrcan.gc.ca/communities-government/buildings/federal/federal-buildings-initiative.cfm.
6. See http://oee.nrcan.gc.ca/communities-government/buildings/federal/about.cfm?attr=28.

9. BRAZIL PUBLIC BENEFIT WIRE-CHARGE MECHANISM

INTRODUCTION

Public benefit funds or wire-charges have been used in many countries after the introduction of power sector reforms.[1] Their function is to collect revenues in an equitable manner to support energy efficiency and other "public interest" programs that might be lost in a restructured utility environment. Brazil is probably the only developing country so far to have introduced a wire-charge mechanism as defined above.[2] Within the Three Country Energy Efficiency Project, this mechanism attracted considerable interest from the Chinese and Indian teams, since the wire-charge has generated substantial funds to be used for energy efficiency and renewable energy investments. In fact, it seems that the Brazilian ESCO industry owes its survival in recent years to a large extent to the wire-charge mechanism.

INSTITUTIONAL ARRANGEMENTS AND RESULTS

Power sector reforms started in Brazil in 1995. Privatized distribution utilities were obliged by clauses in their concession contracts to invest in energy efficiency and R&D activities. Starting in 1998, this was formalized through rulings of ANEEL, the regulatory agency of the power sector, mandating a "wire-charge" of 1 percent of annual utility net revenues which must be used, primarily by the utilities themselves, for the public-benefit investment specified. Distribution utilities that were still state-owned were also required to participate when their concessions were renewed. Generation and transmission companies also had to start contributing to this wire-charge beginning in 2000, but none of those resources were destined for energy efficiency.

The allocation of wire-charge revenues among different programs and types of applications is subject to regulations established by ANEEL, which also approves the project proposals of the utilities for use of these funds and oversees compliance with norms. However, with the exception of some blocks of resources, utilities are responsible for designing and executing all their own programs and projects.

The allocation of wire-charge revenues has changed very significantly since the initial implementation of the wire-charge in 1998, and the Brazilian Congress has since passed several laws that impact the specifics of the wire-charge program. Table CS9.1 summarizes the changes in allocations and shows the gradually diminishing share allocated to energy efficiency (the values are in terms of percent utility receipts, with the total always equal to 1 percent). The 2007 law passed by Congress reinstates the energy efficiency allocation to 0.50 percent, half of which must be spent on energy efficiency measures targeted at low-income households.

In the initial phase, utilities could allocate up to 0.65 percent of the 0.90 percent for energy efficiency to "supply side" measures, thereby reducing their technical and commercial losses. The legislation of 2000 restricted applications to end-use measures, such as HVAC systems in public and commercial buildings, energy efficient motors in industry, and so forth. Utility programs for education on electricity saving and municipal energy management are also eligible. This change was more consistent with the objectives of the wire-charge, since in the newly liberalized environment utilities already have strong incentives to reduce their losses.

Over time, successive rounds of rulemaking by ANEEL gradually restricted utilities' options and came to require maximum cost-benefit

Table CS9.1. Allocation of Wire-Charge Uses in Brazil, 1998–2007 (Percent)

Year	Energy efficiency	R&D[a]	MME/EPE
1998–99	0.90	0.10	—
2000–03	0.50	0.50	—
2004–05	0.50	0.40	0.10
2006	0.25	0.60	0.15
From 2007	0.50	0.40	0.10

Source: Jannuzzi 2005 and Law 11465 of 28/03/2007.
EPE: Energy Planning Company, owned by the Ministry of Mines and Energy (MME).
a. Starting in 2000, half the resources for R&D are managed by the utilities in regulated programs and half go to a national fund called CTEnerg.

ratios (0.80 for most projects, 1.00 for public lighting). Use of wire-charge resources for marketing was eliminated and minimum allocations for different economic sectors were required. These allocations and norms were the same for all utilities, regardless of the large differences in the size and markets of different utilities. At the same time, projects could be extended to more than one year.

Although all projects initially were implemented on a grant basis, in later cycles utilities were allowed to recuperate their energy efficiency expenditures under part of the wire-charge program (currently 50 percent) using performance contracts with the beneficiaries, excluding contracts in the education, municipal, or residential sectors. Part of the returned funds would be used for new energy efficiency projects, and part to reduce electricity rates for consumers.

Table CS9.2 summarizes the investments in end-use programs in the annual cycles since 1998, as well as the estimated energy savings and avoided system demand. The annual cycles do not follow calendar years. The investment values since 2002 are estimates, since official figures have not been disclosed. Still, investments through the wire-charge are about five times larger than investments by PROCEL, the government program for energy efficiency in the electricity sector, amounting to US$70 million during the same time period. The estimates of avoided demand and energy savings are even patchier, since there is little if any systematic verification of the results.

Table CS9.2. Total Investment in Regulated Utility Energy Efficiency Programs in Brazil (1998–2004)

Cycle	No. of utilities	Total investments (US$ million)	% in end-use programs	Avoided demand (MW)	Energy savings (GWh)
1998–99	17	68.3	32	250	754
1999–2000	42	75.9	40	369	994
2000–01	53	35.4	94	n.a.	n.a.
2001–02	60	57.2	99	496	1,498
2002–03[a]	28	39.8	100	n.a.	n.a.
2003–04[b]	40	66.8	100	n.a.	n.a.

Source: Jannuzzi 2005.
n.a. = Not available.
a. Estimated in Jannuzzi (2005), based on data from the major 28 utilities.
b. Estimated in Jannuzzi (2005), based on personal information from the Association of Power Distribution Utilities.

Public lighting, which comprises about 3 percent of annual electricity consumption, has captured most of the resources of the energy efficiency programs of the utilities (see figures CS9.1 and CS9.2). The reasons are the relatively low tariff for street lighting,[3] despite its impact on peak load demand, and the poor payment history of many

Figure CS9.1. Breakdown of Brazilian Utilities' Energy Efficiency Investments by Sector (1998–2003)

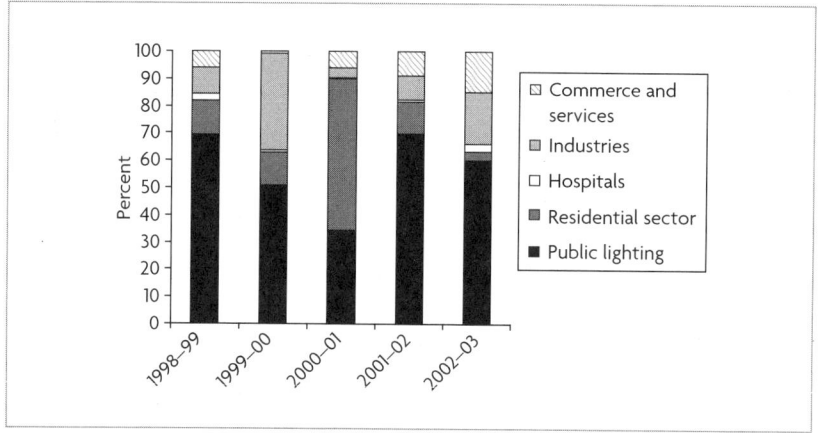

Source: Jannuzzi (2005), based on personal information from the Association of Power Distribution Utilities.

Figure CS9.2. Total Utilities' Energy Efficiency Investments by Sector (1998–2003)

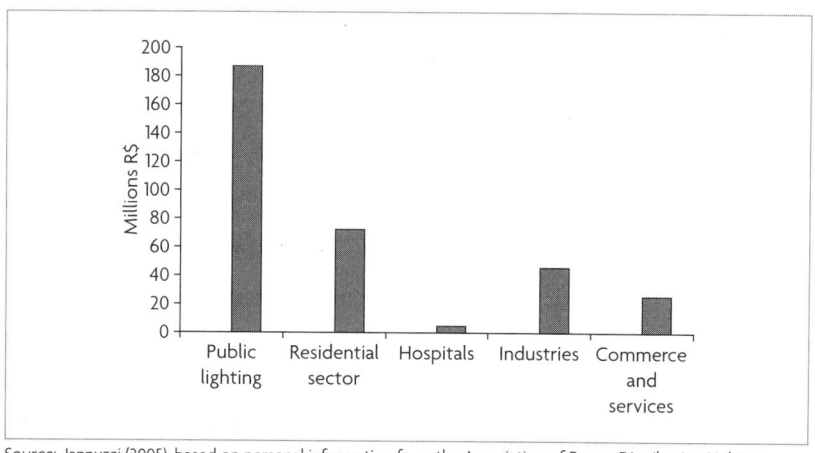

Source: Jannuzzi (2005), based on personal information from the Association of Power Distribution Utilities.

municipalities. From the utilities' perspective, energy efficiency investments were a way to minimize losses from these sales to municipalities. In addition, concessional funds from another federal program can be used for a portion of the investment costs.

The inflow of financial resources through the wire-charge has created an important source of income for some ESCOs and engineering consulting firms. According to a recent survey conducted by ABESCO,[4] Brazilian ESCOs rate the regulated energy efficiency programs as their main funding source. Some of the largest utilities in the country are increasingly outsourcing energy efficiency projects to ESCOs. Utilities decide the types of projects in which they are interested, and ESCOs compete for designing and implementing the projects. During 2002, for example, 117 contracts were signed with ESCOs, representing about 20 percent of the investments in the energy efficiency utilities' regulated programs (see table CS9.3). These particular contracts, however, are not performance contracts but conventional engineering services contracts with remuneration on a cost–plus basis.

ADVANTAGES AND DISADVANTAGES OF WIRE-CHARGE FUNDING

The wire-charge mechanism as implemented in Brazil is an important source of energy efficiency investments funds that would most likely

Table CS9.3. ESCO Contracts with Brazilian Utilities in Energy Efficiency Regulated Programs (2002)

Sector	No. of energy efficiency projects	Investments (million R$)	Average energy savings (%)	Simple pay back (years)
Commercial	35	5.45	18.5	2.4
Public	25	3.67	18.0	1.2
Industrial	57	14.42	19.5	3.0
Total	117	23.54	18.8	2.2

Source: Cited in Jannuzzi (2005): ABESCO 2002, from Jannuzzi, G. M., M. Danella, et al. Metodologia para Avaliação da Aplicação dos Recursos dos Programas de Eficiência Energética. International Energy Initiative. Campinas, p. 10, 2004. (Energy Discussion Paper no. 2.60-01/04.)
Note: US$1.00 = R$2.92 (average for 2002).

not have been available in the privatized power sector without the regulator's enforcement. However, the way it was implemented was far from optimal and has done little to transform the energy efficiency services market. Among the drawbacks of the Brazilian program are the following:

- It is implemented by utilities that have little interest in reducing demand through energy savings, since under existing rates they are likely to lose revenue in most market segments.
- Rigid criteria are used to determine expenditure shares in economic sectors, independent of the size of the utility or the characteristics of the market in its franchise area.
- Little if any (independent) ex post evaluation of costs and benefits of implemented projects has taken place, while there is excessive bureaucracy in the ex ante process.
- The program is highly fragmented with many small and short-term projects.
- Participation of customers or market forces in program design and implementation has been very limited.
- It is not backed by public energy efficiency policies providing strategic guidance on how to maximize social benefits.
- It has not resulted in leveraging of resources or involved commercial financing.
- It has been an important source of revenue and projects for ESCOs, but has done little to prepare them for a more sustainable future based on commercial financing.

The energy efficiency wire-charge could be reformed in several ways to make it a more effective tool to transform the market for energy efficiency in general and also more specifically for energy efficiency retrofit projects. Some ideas are summarized in box CS9.1.

BOX CS9.1. Summary of Some Ideas for Reform of Brazil's Wire-Charge

1. Establish a mid- to long-term programmatic strategy for the wire-charge and utility programs using guidelines for a national energy efficiency policy, following a comprehensive review of the market potential and barriers for energy efficiency in the country.
   a. The guidelines should consider how to achieve better coordination among utilities' programs *and* with other existing mechanisms to support energy efficiency activities, such as PROCEL, CTEnerg (for relevant R&D), the Energy Efficiency Law, and others.
   b. The guidelines should consider complementing utilities' activities with those of independent nonutility agents, who could receive part of the wire-charge funds for implementing energy efficiency programs. Alternative ways to administer energy efficiency programs (for example, the efficiency utility concept adopted in Vermont) could be investigated.
2. Require utilities to show how the wire-charge resource is being leveraged and provide incentives for leveraging.
   a. Develop possibilities to use resources to leverage performance contracting by ESCOs, approaching more closely a normal commercial financing model (including loans) and achieving higher leverage of resources.
3. Enable pooling of resources where cost effectiveness can be improved.
4. Simplify protocols for ex ante project approval and develop differentiated program design rules according to utility size and market characteristics.
5. Introduce independent ex post program evaluations. Administrative fees might be collected for independent evaluation and monitoring.

Source: Based on Jannuzzi (2005).

NOTES

1. Unless otherwise noted, information in this case study is heavily drawn from Jannuzzi (2005).
2. In Thailand, an electricity tariff adjustment provided resources for the Electricity Generating Authority of Thailand's DSM program, the precursor of the Energy Conservation Promotion Fund which is funded through a petroleum products surcharge; see http://www.eppo.go.th/encon/encon-fund00.html.
3. For example, the low-voltage energy tariff of CEMIG (a power utility, headquartered in Belo Horizonte, Minas Gerais) for public lighting is R$204/MWh, compared with R$379/MWh for low-voltage business customers. A medium-voltage customer with a load profile similar to public lighting would pay about R$280–285/MWh.
4. ABESCO (2005a, 2005b).

10. SRI LANKA DSM: USING THE UTILITY BILL AS A LOAN REPAYMENT MECHANISM

INTRODUCTION

The Ceylon Electricity Board (CEB) is a vertically integrated public utility responsible for electricity generation, transmission, and distribution in Sri Lanka. In order to cope with a power crisis in the mid-1990s, CEB initiated a DSM program with the objective to increase the efficiency in the use of electricity, resulting in lower electricity bills for the customers, the mitigation of CEB energy deficits in the short term, and deferment of CEB investment in new capacity over the long term.

This case study demonstrates how a load management program and energy efficiency benefits can coincide in certain situations.[1] In this case, program benefits (up to 2000) included 74 MW of demand savings and 110 GWh per year of energy savings. The system load factor improved from 57 to 60 percent due to DSM. The program received wide public acceptance and transformed the compact fluorescent lamp (CFL) market in Sri Lanka. The number of CFLs purchased by customers from outside of the program due to program publicity (1,235,000) far outweighed the number of CFLs purchased by the program participants (261,000) as presented in the program's evaluation of 2001.

This program structure utilizes monthly utility bills as a mechanism to collect loan repayments for energy efficiency investments. In order to make this possible, the legal implications of the contractual arrangements between the utility and its customers need to be considered carefully.

PROGRAM DESCRIPTION

The program was implemented by the DSM branch of the CEB, and implementation arrangements were progressively refined from the initial pilot in 1994 to the third stage, which ended in 2001. CEB partners in this program included the Lanka Electricity Company (a private distribution company) as well as the Energy Conservation Fund, which was responsible for implementing the program in the public sector. The program aims to reduce both peak load requirement and energy use through sale of more energy-efficient lamps to utility customers. A schematic is shown in figure CS10.1.

The CEB maintains a list of approved lighting models, costs, and dealers, and this list is updated regularly. The program is marketed by all the listed suppliers in a competitive environment. Listed suppliers offer quality standards for their lamps and provide a two-year warranty as well. In order to be approved by the CEB, suppliers need to have their products tested independently. The DSM branch provides assistance in arranging this testing.

The program is promoted through awareness campaigns funded by the CEB. The Government has shown its support to the program by providing a waiver to cover the import taxes and other duties. The CEB also plays a key role in overcoming the high first-cost barrier by offering an interest-free credit facility to the participating customers. Customers are required to sign an agreement with the CEB to pay for the lamps (limit of four lamps per customer) in 12 monthly installments. The monthly payments are collected as part of the monthly electricity bill. Upon signing the agreement, the customers would collect the lamps from one of the participating dealers and the dealer in turn would be reimbursed by the CEB for the full cost of the lamps.

Figure CS10.1. Sri Lanka DSM Project: Institutional Arrangements

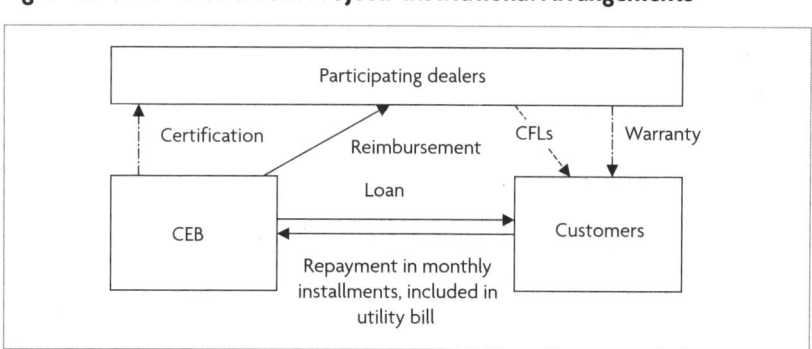

Source: Authors.

A key feature of this program is the mechanism utilized for collecting repayments from participating customers. Although this arrangement provides additional comfort regarding repayments since it is linked to the bill, actual leverage on the customer is limited because disconnection due to nonpayment of loan installments is illegal.

The utility has one agreement with participants as an electricity customer and a separate agreement for participating in the program and making the monthly payments. Customer accountability for each of the agreements needs to be kept distinct. For this reason, the customer cannot be disconnected for nonpayment of the monthly installments for the energy efficiency loan—even if the loan is collected as part of the monthly electricity bill—as long as the electricity bill is paid. In the case of the CEB DSM program, bill payments from customers first were applied to loan installments and any remaining balance to the electricity bill. In this manner, the utility was effectively utilizing the bill as a channel for collecting customer repayments.

NOTE

1. Sri Lanka and many other developing countries often have a power system load profile that is characterized by a sharp evening peak, primarily attributable to electric lighting. In a power sector where both capacity and energy are scarce, adopting an efficient lighting program brings significant savings in both capacity and energy.

11. DONGYING SHENGDONG EMC WASTE GAS POWER PROJECTS

INTRODUCTION

In a short time, the Dongying Shengdong Energy Management Company (EMC or ESCO) has established a thriving business that involves installing and operating power generation equipment for industrial clients. The company's clients provide waste gas (free of charge) to act as a feedstock, and buy the on-site electricity from the EMC at a marked-down price compared with grid electricity. Based on revenue-sharing arrangements outlined in a standard 10-year contract, the EMC can recoup its main capital investment in about two years, and then maintain a profitable operation and maintenance relationship for the balance of the long-term contract. The company has combined its technical knowledge and business savvy to generate substantial profits, while at the same time benefiting customers. This type of chauffage arrangement has a strong future, especially considering the amount of unused waste gas available in China and tight energy supply.

COMPANY AND BUSINESS BACKGROUND

The Shengli Oil Machinery Company is a well-established manufacturer of power stations that are capable of burning waste gases, such as coal mine methane or waste gas from coking plants. In 2000, several of its managers and employees set up the Dongying Shengdong Machinery Company Ltd. (referred to below as the Dongying Shengdong EMC).

This company's business strategy is relatively straightforward. It seeks industrial clients that can provide it with waste gas, at no or little cost, for use as a feedstock in power stations. Once a client is found, the EMC utilizes mature technology from the Shengli Oil Machinery Company, installs the equipment on site at the client's facility, and burns the waste gases in a power station to produce electricity. The

Figure CS11.1. Dongying Shengdong EMC Ownership and Business Arrangements

```
                    Shengli Oil
                    Machinery Co.
                         ↕                    Loan
Ownership                                   repayment         Bank
(through         Profit    Equipment         ─────→
employees/                  supply                              ↕
managers)                                  Bank loan with    Investment &
                 Dongying Shengdong      guarantee from I&G  Guarantee Co.
                        EMC
                                          Waste gases
                                          Payment for electricity
                          Electricity
                          generation
    Customers             equipment,         Customer
                          electricity
```

Source: Authors.

electricity is then sold to the client at a marked-down price. The EMC finances the investment and recoups it through this sale.

By early 2004, the company had installed 25 separate power stations, with a combined capacity of around 45 MW, and was working on 13 more power stations, with a combined capacity of an additional 40 MW. The structure and contractual arrangements of a typical project are described in figure CS11.1.

SAMPLE PROJECT: A 1.9 MW POWER STATION AT A COKING PLANT

Technical content and responsibilities. In this project the Dongying Shengdong EMC financed and installed a power station at a coking plant. The power station included six of Shengli Oil Machinery Company's gas-burning units. The coking company client agreed to provide waste gas (of a specified quality) free of charge to the EMC, which used the gas in the power station to produce electricity, which was then sold to the client. The Dongying Shengdong EMC was responsible for the following:

- providing (but retaining ownership of) the power equipment itself
- providing design blueprints

- transporting the equipment to the site
- overseeing the installation of the power station
- sending personnel to handle the maintenance, testing, management, and operation of the power station
- being subject to overall management by the client (that is, being subject to safety inspections conducted by the client, providing power based on the demands of the client, respecting the client's onsite code of conduct, and so forth)

In addition to providing free waste gas and cooling water, the client was responsible for providing the space for the power station, all other ancillary equipment, utility connections, installation of the equipment, and provision of financial guarantees to the EMC.

Investment and financing approach. The project required a total of Y 7.4 million (about US$0.9 million) of investment on the part of the Dongying Shengdong EMC. It obtained a Y 7 million loan from a commercial bank, 90 percent of which was guaranteed through the WB/GEF loan guarantee fund (see Case Study 1). The remainder of the investment came from the EMC's own internal funding. I&G, the guarantee company, considered the EMC's many power stations suitable collateral and accepted them as counter-guarantees. The Y 7 million loan had a one-year term. As the project generated Y 3.98 million in annual revenue for the EMC from electricity sales to the client (see below), a good percentage of the loan was paid back entirely from project cash flow. The rest was paid back using other resources.

Repayment agreement with the customer. It was estimated that the project would generate Y 6.127 million in revenue each year, by partially replacing grid electricity with electricity generated onsite. The contract between the EMC and the client stipulated that these savings in electricity costs compared to grid supply would be shared over a period of 10 years, with 65 percent going to the EMC and 35 percent going to the client. The host company, which no longer had to pay the grid company for this electricity, instead paid 65 percent of what it would otherwise have paid to the EMC, resulting in Y 3.98 million in annual project revenue for the EMC.

The contract stipulated that the price of electricity charged from onsite generation would be lower than the grid-supplied tariff (Y 0.45 per kWh at time of contract signing with the EMC), and that it would be adjusted in line with tariff adjustments for industrial customers of the local grid company. However, the minimum price would be Y 0.30 per kWh (including tax).

In addition to maintaining ownership of the power station, thus allowing the EMC to remove the equipment and use it in other projects in case of permanent default on the client's part, the EMC's contract also includes provisions to ensure that the client pays its bills on time. For instance, if the client failed to fulfill its contractual obligations, it would be required to pay the equivalent of three months of electricity bills to the EMC. In addition, the client would also be contractually obligated to cover all fees associated with disassembling and transporting the equipment away from the site.

On the other hand, the contract also states that if power generation halts due to the EMC not fulfilling its obligations, the EMC is required to pay three month's worth of electricity fees to the client as compensation (in such a case, fees would be calculated based on the power station's average monthly electricity production).

LESSONS LEARNED

By combining business acumen, technical knowledge, and the availability of a specific and reliable technology, the EMC has been able to establish a rapidly expanding business of providing chauffage services, which are also quite beneficial to its clients.

The main concern of clients is to ensure that their production processes are not interrupted. If the power station is built and provides cost savings, this is beneficial, but the client also has its grid connection for power supply in the event of problems. With the investment requirement of the clients being small, as the EMC finances the power plant and its installation, risks to clients are relatively low.

12. IQARA ENERGY SERVICES IN BRAZIL

INTRODUCTION

Iqara Energy Services (IES)[1] is a wholly owned subsidiary of Iqara (London), itself a wholly owned subsidiary of British Gas, which creates specific units for different business opportunities in several countries outside the owner's core regulated business of distributing natural gas. British Gas is the majority owner (60 percent) of Comgás, the natural gas distribution utility serving metropolitan São Paulo, and also has reserves of Bolivian gas and a share of the Bolivia-Brazil gas pipeline.

IES is an entirely private sector, market-driven initiative. Its primary objective is not energy efficiency, but to create new markets for natural gas. However, the strategy chosen to achieve this end usually results in energy efficiency improvements.

Attracting customers is a major challenge in many developing countries beginning to expand their natural gas infrastructure. In many sectors existing fuel use is small, and simple fuel substitution will generate only a tiny increment in gas sales volume. Achieving significant sales would require substituting electricity in some end-uses. In other sectors with somewhat larger fuel use, simple substitution may not be compellingly attractive on price terms alone—other potential advantages offered by natural gas must be exploited. In both types of cases, substantial investments may be required and the client may not be willing or able to finance them.

IES provides 100 percent financing, with an integrated package of services that includes a guaranteed level of cost savings compared to existing energy costs. The main focus is on small-scale cogeneration (up to 5 MW, but usually much smaller) packaged with absorption chillers, though other, less efficient configurations are also implemented.

Iqara Energy Services can be viewed both as an example of a DSM operation or an ESCO project aggregator. It tends to be more like the latter because (i) it does not do the project engineering, but concentrates on the financial and contractual aspects of the project; and (ii) it does not restrict its activities to the franchise area of the distribution utility.

Since it began operation in January 2003, IES has been quite successful in developing small cogeneration projects in a market that had been extremely difficult to open. Based on this favorable experience, BG India Energy Services Private Limited was created in 2004 to develop cogeneration facilities in India. Based in Surat, the commercial hub of Gujarat, it replicates the basic business model established by IES in Brazil and has already implemented some projects in India.

INSTITUTIONAL ARRANGEMENTS AND RESULTS

IES offers a comprehensive package, which includes 100 percent financing of the project and the integrated services to implement it, including project design; acquisition and installation of the equipment; subsequent operation and maintenance (O&M), and management of the energy supply contracts both for electricity and natural gas. It guarantees that the cost of the energy services provided by natural gas will be lower than the equivalent services provided by the preexisting supply configuration, following this equation:

$Y = X - $ (**guaranteed percent of X**), where:
 $X = \Sigma$ of current costs (electricity cost + fuel bill + O&M)
 $Y = \Sigma$ of costs with IES (equipment + natural gas
  + backup/compl. power + O&M)

The reduction in costs is typically in the range of 3–10 percent. This package, which minimizes customers' costs besides taking the investment off their balance sheets, is a kind of chauffage performance contract. The contracts are quite long term (15 years length is common), especially by Brazilian standards. During this period, IES maintains ownership of the installed equipment. Figure CS12.1 summarizes Iqara's business model.

Figure CS12.1. Iqara's Business Model

```
Equipment manufacturers ─┐
                         ├─ Subcontracts ──► Iqara Energy Services ◄── Monthly payments ── Industrial/commercial client
Engineering firms/ESCOs ─┘                                          ── Design solution ──►
                                                                       Finance investment
                                                                       Install equipment
                                                                       Operate/maintain
                                                                       Manage energy contracts
                                                                       Supply natural gas
```

Source: Authors.

The basic market segments targeted by IES in its start-up phase have the following characteristics:

- The market consists of medium-size energy consumers (up to 5 MW of demand, usually much less), paying the A4 tariff (connected at 2.3–25 kV). These consumers typically pay 40–50 percent more per kWh than large consumers.

- The consumers have a potential to substitute electric chillers with absorption chillers. The former may need replacing or there may be a new installation/building.

- The profile of electrical and thermal demand is such that with the implementation of the cogeneration project all the electricity produced is consumed on site by the client. This avoids complex negotiations to sell electricity to the grid at a much lower price than the avoided cost of buying electricity.

- The consumers give a high priority to the reliability of the energy services that were being provided by the electricity distribution utility.

In the beginning, IES restricted its marketing to the region served by Comgás, also owned by British Gas. However, in 2004 it began to sell its services outside the franchise area, with a project in a prestigious hotel in Rio de Janeiro.

Most of the projects implemented by IES are cogeneration projects, where the thermal base is heat for an absorption air conditioning system (see table CS12.1). However, there are several projects that do not involve cogeneration. Three are direct-fired absorption chiller projects and one is a peak load generating plant. None of these projects can be characterized as improving energy efficiency, though they provide cost savings and reliability benefits for the client and certainly increase sales of natural gas. It is possible that the direct-fired projects could be upgraded later to be cogeneration projects, but this depends on the technology chosen for the absorption units. The projects implemented until now are smaller, usually much smaller, than the 5 MW upper limit set for the first years of IES business development. No estimate is available for the total investment in the projects shown, but it appears to have been US$10–12 million.

IES' business is more about selling a set of economic solutions using natural gas than energy efficiency as such. On the other hand, most ESCOs in Brazil do not restrict themselves to projects involving increased energy efficiency, but also provide solutions that reduce

Table CS12.1. Iqara Projects

Sector	Project	Power capacity (kW)	Thermal output
Super-markets	Sonda Penha	395	140 TR
	Sonda Santo Amaro	375	110 TR
	Sonda São Bernardo Campos	790	280 TR
Hotels	Sofitel	300	100 TR + hot water
	Caesar Park Guarulhos	395	135 TR
	Caesar Park Ipanema	0	210 TR
Shopping centers	Taubaté	0	560 TR
	Taboão	2,855 (800 only for peak)	1600 TR
	Mega Pólo	0	900 TR
	Osasco	1,300 (peak hour)	0
Industry	Inapel Embalagens	1,120	170 TR + 5 t/hr steam

Source: Authors, based on personal communication from IES.
TR: Tons of refrigeration.

costs or increase reliability. Given the severe differential of peak versus off-peak power in Brazil (an A4 consumer pays 8–10 times more for peak power at the same load factor), solutions that reduce peak hour demand and consumption are especially attractive.

IES does not undertake the plant engineering, installation, maintenance, and operation. This work is outsourced. IES is not tied to any particular equipment supplier or provider of engineering services, though in an initial phase almost all projects were implemented in a partnership with one company.

IES reports plans to expand substantially in the coming years, with a goal of completing 600 projects and US$300 million by 2010. Although the emphasis will continue to be on small projects, Iqara is also pursuing the possibility of implementing somewhat larger projects in industries that require capacities of 10–20 MW.

ADVANTAGES AND DISADVANTAGES OF THIS MODEL

Given the short time of its existence, it is difficult to evaluate the success of IES's strategy. In its first two years, IES developed, by Brazilian standards, an impressive portfolio of small cogeneration projects[2]. Sales slowed in 2005—which was a difficult year for vendors of natural gas cogeneration systems due to the political crisis in Bolivia and uncertainties about policy for gas interruptibility. However, new projects were being contracted and implemented in 2006.

The integrated package offered by IES is in principle potentially attractive to many clients. However, the savings guaranteed are relatively small and the length of contracts has been quite long.

Controlling costs is a key challenge in this business. Marketing costs are typically high and only a small fraction of initial contacts result in projects. Furthermore, since the projects are small, development and transaction costs must be controlled. At the same time, care must be taken in developing cogeneration projects not to adopt a too-standardized "cookie cutter" approach which may not work for many clients.

A key aspect of IES's business model is that all financing is with the firm's equity. There is no debt finance. So far, debt financing has been considered too cumbersome, with delays, uncertainties, and high transaction costs.

The cost of the package is increased by the fact that IES assumes all risks, including changes in the relative prices of energy inputs (basically electricity and natural gas), which are quite unpredictable over the long term of the contract. As discussed elsewhere, the best agent to assume energy price risk in energy efficiency projects is usually the consumer. If the price of energy increases, the project is more profitable. If the price decreases, the consumer gains a windfall from his total consumption. In the case of substituting electricity with natural gas, the situation is somewhat different, but the relative prices remain crucial. Price risk coverage may be interesting to some consumers, despite its implied expense, but it requires a large and sophisticated backer.

From the perspective of considering IES as a model for an energy efficiency business, there is an inherent tension between the objective to simply sell more gas and energy efficiency objectives. As discussed above, direct-fired absorption systems are not more efficient than modern electric chillers (assuming for the comparison that the gas is used in combined-cycle central stations). Rather, they satisfy the goal of selling gas with a much smaller investment than a cogeneration system.

LESSONS LEARNED

IES followed a strategy similar in many ways to that pioneered by Enron Energy Services (EES) in Brazil in the late 1990s. However, EES failed to close a single deal, despite having established a large and sophisticated operation. IES's more successful startup appeared to be due in part to a more focused strategy. Whereas EES sought projects of all kinds with business consumers, IES defined a niche that is quite large (potentially several thousand installations) and specialized in it.

However, even with strong financial backing, technical credibility, and provision of 100 percent financing to clients, marketing has not been easy. Potential clients may regard the service as expensive (the guaranteed saving is actually quite small) and the term of the contract too long. They may not fully understand the difficulties and risks of implementing such projects if they were to attempt them alone.

Controlling fixed and transaction costs are crucial for a company developing and financing small projects. Like EES before it, IES may have started with a structure that was too elaborate and costly. It seems to have taken steps to reduce these costs, including the merging of IES with Iqara Natural Gas, a business unit created to work in the market for compressed natural gas.

Incorporating some debt financing could in principle reduce costs and hence permit larger guaranteed savings (or shorten the term of contracts). However, competitive and efficient local debt financing arrangements for such medium-sized projects remain difficult to obtain in Brazil.

NOTES

1. See http://www.iqaraenergy.com.br/index.asp.
2. It is worth noting that only about 20–30 small cogeneration projects using natural gas have so far been implemented in the country.

ENERGY EFFICIENCY FINANCE CASE STUDIES 257

13. INDIA CAPACITOR LEASING

INTRODUCTION

The Indian electricity system is often characterized by poor levels of power factor, particularly in the distribution system. Poor power factor results in higher system losses and consequently low technical and business efficiency. The existing tariff structure for high tension (HT) consumers in India typically imposes a penalty on customers who do not maintain a 0.8 power factor. Ideally, utilities would like to maintain a 0.9 or higher power factor. Despite the tariff penalties on consumers, utilities find it difficult to maintain a high power factor and are looking for ways to increase the power factor and reduce system-wide technical losses.

The following case study presents an innovative leasing approach to address the problem of low power factor among industrial consumers. The Ahmedabad Electric Company (AEC) implemented such a capacitor leasing program in association with a vendor ESCO, Saha Sprague Limited (SSL). SSL received debt financing from IREDA to help in the implementation of this project (as presented in Case study CS4). The structure described in figure CS13.1 has allowed the utility to achieve its power factor goals in partnership with an ESCO.

DESCRIPTION

Under this program, the ESCO (SSL) helps the utility to achieve a high power factor under a performance guarantee. SSL is required to provide a performance guarantee to the utility (AEC) that the equipment supplied would perform at a minimum of 90 percent of the designed value. The capacitor leasing program has been implemented based on a framework of three sets of contracts.

First, the vendor ESCO SSL signed a memorandum of understanding (MOU) with the utility AEC for improving reactive power demand in

Figure CS13.1. Institutional Arrangements in a Capacitor Leasing Program in India

```
                Loan
   ┌────────┐ ────────► ┌──────────────┐   Equipment   ┌──────────┐
   │ IREDA  │           │ Saha Sprague │    lease      │ End user │
   │        │ ◄──────── │  Ltd. (SSL)  │ ────────────► │          │
   └────────┘ Repayment └──────────────┘               └──────────┘
         First        Lease rental      Electricity    Electricity bill
         charge       fee                              Lease rental
                                          MOU
                      ┌──────────────┐                 ┌──────────┐
                      │   Escrow     │ ◄────────────── │   AEC    │
                      │   account    │ Lease rental    │          │
                      └──────────────┘ minus admin. cost└──────────┘
```

Source: Authors.

selected supply areas with a target of achieving a power factor of 0.95. Under the MOU, SSL will supply, install, own, and maintain capacitors at consumer premises on lease. At the end of the lease period, the ownership of the capacitors is transferred to the lessee upon payment of a nominal amount. The second set of contracts is a tripartite agreement between each participating consumer, AEC, and SSL. Third, IREDA and SSL signed a loan agreement through which SSL received debt financing. This helped to finance the capital cost of the equipment installed at consumer premises.

Lease payments are collected from the consumers by AEC through the billing system and are transferred to SSL after deducting 3 percent toward administrative costs. SSL then makes the loan repayments back to IREDA. If a consumer fails to pay the lease rental, SSL is entitled to remove the equipment supplied and also collect the outstanding amount. If the performance of the equipment is not satisfactory, AEC can terminate the MOU with SSL. Similarly, if AEC does not transfer the payments, SSL has the ability to terminate the agreement.

IREDA has put an interesting repayment security arrangement in place to ensure that the debt provided to SSL is serviced on time. The repayment security is provided through the creation of an escrow account that serves as a channel for the AEC to transfer the rental collections to SSL. Under the loan agreement by SSL with IREDA,

IREDA has first charge on the escrow account and is insulated from repayment risks as payments flow directly through the escrow account.

LESSONS LEARNED

- At the core of this transaction is the performance guarantee that the ESCO is able to offer to the utility. In this case, the ESCO is offering an energy efficiency service that can be easily monitored. Power factor improvements are relatively less complicated to verify given the metering technologies available and fewer external factors that can influence power factor reductions.
- The utility's MOU with the ESCO provides the consumers a high level of comfort regarding the company and the product that is to be installed. The leasing arrangement is also attractive to the consumers as compared to the option of cash purchase.
- The escrow account serves as an effective mechanism to secure loan repayments to the financial institution.
- In addition, the agreement between the ESCO and the utility provides the ESCO with a higher level of comfort with respect to the consumer lease payments, as compared to a situation where the ESCO is directly dealing with consumers.

BIBLIOGRAPHY

ABESCO. 2005a. "Análise dos Resultados da Pesquisa das Empresas de Serviços de Eficiência Energética no Brasil." February. São Paulo.

———. 2005b. "Resumo das Respostas ao Questionário para Empresas de Serviços de Eficiência Energética." January. São Paulo.

Acemoglu, Daron, and Simon Johnson. 2005. "Unbundling Institutions." *Journal of Political Economy* 113(5): 949–995.

Aghion, Philippe, and Patrick Bolton. 1997. "A Trickle-Down Theory of Growth and Development with Debt Overhang." *Review of Economic Studies* 64: 151–72.

Allen, Franklin, Rajesh Chakrabarti, Sankar De, Jun "QJ" Qian, and Meijun Qian. 2006. "Financing Firms in India." Unpublished paper. At: http://www.darden.virginia.edu/em/PDFs/Allen_Franklin.pdf.

Allen, Franklin, Jun Qian, and Meijun Qian. 2005. "Law, Finance, and Economic Growth in China." *Journal of Financial Economics* 77(1): 57–116.

Amram, Martha, and Nalin Kulatilaka. 1999. *Real Options: Managing Strategic Investment in an Uncertain World*. Boston, MA: Harvard Business School Press.

Banerjee, Abhijit, and Andrew Newman. 1993. "Occupational Choice and the Process of Development." *Journal of Political Economy* 101: 274–98.

Beck, Thorsten, Asli Demirguc-Kunt, Luc Laeven, and Ross Levine. 2005. "Finance, Firm Size, and Growth." Policy Research Working Paper 3485. World Bank, Washington, DC.

Bernstein, Peter L. 1992. *Capital Ideas*. New York: The Free Press.

Coase, Ronald. 1937. "The Nature of the Firm." *Economica* 4: 386.

———. 1960. "The Problem of Social Cost." *Journal of Law and Economics* 3: 1.

Crestar. 2005. "Designing Financial Structures and Financing Instruments for Energy Efficiency Projects in India." Consultant Report for the 3CEE project. Mumbai. http://www.3countryee.org/public/EEStructuresInstrumentsIndia.pdf

Energy Sector Management Assistance Program (ESMAP). 2000. "Operating Utility DSM Programs in a Restructuring Electricity Sector—Summary." Workshop Proceedings. ESMAP, Washington, DC.

Fernandez, Ana Margarida, and Aart Kraay. 2005. "Property Rights Institutions, Contracting Institutions, and Growth in South Asia: Macro and Micro Evidence." The World Bank. Background paper prepared for the SAARC Business Leaders' Conclave: South Asia Regional Integration and Growth, New Delhi, November 17–18.

Galor, Oded, and Joseph Zeira. 1993. "Income Distribution and Macroeconomics." *Review of Economic Studies* 60(1): 35–52.

Greenwood, Jeremy, and Boyan Jovanovic. 1990. "Financial Development, Growth, and the Distribution of Income." *Journal of Political Economy* 98(5): 1076–1107.

Housing and Urban Development Foundation (HUDF). 2002. "Improving Energy Efficiency in Residential and Public Buildings in Lithuania: The Energy Efficiency Housing Pilot Project." At: http://www.munee.org/go.idecs?i=85.

International Energy Agency. 2006a. *Light's Labour's Lost—Policies for Energy-efficient Lighting*. Paris, France: International Energy Agency.

———. 2006b. *World Energy Outlook 2006*. Paris, France: International Energy Agency.

International Panel on Climate Change. 2007. *Climate Change 2007—The Physical Science Basis*. Contribution of Working Group I to the Fourth Assessment Report of the IPCC. At: http://ipcc-wg1.ucar.edu/wg1/wg1-report.html.

Jannuzzi, Gilberto M. 2005. "Energy Efficiency and R&D Activities in Brazil: Experiences from the Wirecharge Mechanism (1998–2004)." Report for the 3CEE project, June 2005. http://3countryee.org/public/WirechargeMechanismBrazil.pdf.

La Porta, Rafael, Florencio Lopez-de-Silanes, Andrei Shleifer, and Robert W. Vishny. 1998. "Law and Finance." *Journal of Political Economy* 106(6): 1113–1155.

———. 1999. "The Quality of Government." *Journal of Law, Economics, and Organization* 15(1): 222–279.

Levine, Ross. 1997. "Financial Development and Economic Growth: Views and Agenda." *Journal of Economic Literature* 35(2): 688–726.

Magill, Michael, and Martine Quinzi. 1997. *Theory of Incomplete Markets: Volume I.* Cambridge, MA and London: The MIT Press.

Markusen, Ann. 1996. "Sticky Places in Slippery Space: A Typology of Industrial Districts." *Economic Geography* 72(2): 294–314.

Ménard, Claude, and Bertrand du Marais. 2006, "Can We Rank Legal Systems According to Their Economic Efficiency?," in P. Nobel and M. Gets (eds.), *New Frontiers of Law and Economics*, Zürich, Schulthess, 7–27.

Mills, Evan. 2003. "Risk Transfer via Energy-Savings Insurance." *Energy Policy* 31: 273–281.

Murtishaw, Scott and Jayant Sathaye. 2006. "Quantifying the Effect of the Principal-Agent Problem on US Residential Energy Use." LBL Report 59773. Berkeley, CA. At http://ies.lbl.gov/iespubs/59773Rev.pdf

Pacala, Stephen, and Robert Socolow. 2004. "Stabilization Wedges: Solving the Climate Problem for the Next 50 Years with Current Technologies." *Science* 305(5686): 968–972. At: http://carbonsequestration.us/Papers-presentations/ htm/Pacala-Socolow-ScienceMag-Aug2004.pdf.

Schipper, Lee, Celine Marie-Lilliu, and Roger Gorham. 2000. *Flexing the Link between Transport and Greenhouse Gas Emissions: A Path for the World Bank.* Paris: International Energy Agency.

Shi, Xiaoyu. 2007. "Financial Cost Effectiveness of Energy Efficiency Investments—A Summary Study." Unpublished discussion paper. World Bank, Washington, DC.

Stern, Nicholas. 2006. *The Economics of Climate Change – The Stern Review.* Cambridge: Cambridge University Press. At: http://www.hm-treasury.gov.uk/independent_reviews/stern_review_economics_climate_change/sternreview_index.cfm.

Three Country Energy Efficiency Project. 2006. Country reports on Brazil, China, and India. At: http://3countryee.org/reports.htm.

Brazil Country Report. August 2006, at: http://3countryee.org/Reports/Brazil_3CEE_Report.pdf

China Country Report. May 2006, at: http://3countryee.org/Reports/Draft CountryReportChina.pdf

India Country Report, May 2006, at: http://3countryee.org/Reports/Draft CountryReportIndia.pdf

Ürge-Vorsatz, Diana, Pierre Langlois, and Silvia Rezessy. 2004. "Why Hungary? Lessons Learned from the Success of the Hungarian ESCO Industry." Summer Study in Buildings, Asilomar, California, ACEEE, 6: 6-345 to 6-356. At: http://www.econolerint.com/en/PDF/HUNGARY_ESCO_final.pdf.

Ward, William A., James B. London, and Robert P. Taylor. 1994, reprinted 2006. "Overview of Case Studies and Economic Analysis," Part 1 of Report Number 4 of *China: Issues and Options in Greenhouse Gas Emissions Control*. Washington, DC: The World Bank. Reprinted by the Clemson University Center for International Trade in cooperation with the World Bank and available on-line at: http://business.clemson.edu/cit/documents/China%20Energy%20Part%201.pdf.

Williamson, Oliver E. 1985. *The Economic Institutions of Capitalism: Firms, Markets, Relational Contracting*. New York and London: The Free Press.

World Bank. 2002. *Lithuania Energy Efficiency and Housing Pilot Project*. Implementation Completion Report No. 22868. Washington, DC: World Bank.

———. 2004. "World Bank GEF Energy Efficiency—Portfolio Review and Practitioners' Handbook." Thematic Discussion Paper. World Bank, Washington, DC. http://siteresources.worldbank.org/INTCC/812001-1110807496989/20480590/WBGEFEnergyEfficiencyHandbook2004.pdf.

———. 2006a. *Improving Lives: World Bank Group Progress on Renewable Energy and Energy Efficiency Fiscal Year 2006*. World Bank, Washington, DC. http://siteresources.worldbank.org/EXTENERGY/Resources/3368051157034157-861/Improving_Lives_Low_Res.pdf.

———. 2006b. "Integrating Energy Efficiency into World Bank Operations." Good Practice Note. World Bank, Washington, DC.

———. 2006c. "Doing Business Database." http//www.doingbusiness.org.

World Bank/OED. 2005. *OED Review of Bank Lending for Lines of Credit*. Operations Evaluation Department. Report No. 31131, World Bank, Washington, DC.

APPENDIX. GLOSSARY OF SELECTED TERMS IN NEW INSTITUTIONAL ECONOMICS (NIE) THAT RELATE TO ENERGY EFFICIENCY FINANCE

The following glossary makes liberal use of resources provided online by the Institute for Development Studies at Sussex University ("Livelihoods Connect") in the United Kingdom, the Lectric Law Library, the International Society for New Institutional Economics, and the Ronald Coase Institute, for which we express gratitude. Quotation and citation are provided where appropriate.

Contract: "An agreement between two or more parties which creates obligations to do or not do the specific things that are the subject of that agreement.... This term, in its more extensive sense, includes every description of agreement, or obligation, whereby one party becomes bound to another to pay a sum of money, or to do or omit to do a certain act; or, a contract is an act which contains a perfect obligation. In its more confined sense, it is an agreement between two or more persons, concerning something to be done, whereby both parties are bound to each other, or one is bound to the other. Blackstone defines it to be an agreement, upon a sufficient consideration, to do or not to do a particular thing. A contract has also been defined to be a compact between two or more persons."

Contracts may be short-term (spot, immediate) or long-term (dynamic); and they may be formal (written) or informal (spoken or implied). The weaker the countries' formal contracting institutions (that is, the institutional environment), the greater will be the dominance of short-term or spot transactions. Such short transactions do not rely upon more complex institutional arrangements involved in financing or in other forms of future commitments (long-term contracts) that increase the scope for and the need to protect against opportunism.

Source: Quoted section is from "The 'Lectric Law Library's Legal Lexicon on CONTRACT." http://www.lectlaw.com/def/c123.htm. Remainder is by the authors of this report.

Contracting forms in East and West: In general, institutional arrangements for long-term (dynamic) contracts in Western (Anglo-American, in particular) societies (and institutional environments) tend toward formality and a related legalistic search for ex ante completeness. On the other hand, institutional arrangements for long-term contracts in East Asian (Japanese and Chinese in particular) institutional environments tend to combine formal contract models with informal, self-sustaining agreements that are sustained by the value of future relationships, and to focus more on the ex ante resource commitments of building relationships that constrain potential ex post transaction costs that also are more broadly defined (in East Asian as compared to the Western societies) to include general long-term business interests.

Source: The authors.

Enforcement mechanisms are the means by which parties to a contract are induced to fulfill commitments made or implied via that contract and/or via related explicit or implied obligations. Thus, they are essential elements of incentives and of institutional arrangements.

The analysis of "laws" absent a parallel analysis of enforcement mechanisms can lead to erroneous conclusions about the quality of a country's legal and economic institutions. Indeed, "systems of men" (as opposed to "systems of law") often will have a large number and range of laws on the books. This range of laws can give corrupt administrations additional scope for inappropriate implementation by providing a wide range of formal rules that may be selectively enforced in favor of friends and against enemies.

Informal institutions of contracting (see institutions) may rely upon social sanctions such as shunning, ridicule, or even vigilantism for enforcement. Formal institutions of contracting and contract enforcement, on the other hand, usually rely upon actions by courts (or duly constituted arbitrators such as the International Chamber of Commerce) and/or government agencies to adjudicate and enforce agreements made by private parties.

Source: The authors.

"**Institutions** are defined as the formal or informal rules governing peoples' and organizations' behaviour. Institutions, 'the rules of the game', are distinguished from organizations which, along with individuals, are considered as 'players in the game'. *Formal institutions* (such as by-laws, national laws, policies, the national constitution, and international laws and treaties) are part of the institutional environment and distinct from institutional arrangements. For *informal institutions* (such as social customs and conventions), the distinction may not always be so clear. However, widespread acceptance of particular institutional arrangements as the norm can mean that in effect they become part of the institutional environment."

Informal institutions can be weak in facilitating contracting that spans more than one familial or social/ethnic group and, thus, can limit the ability to exploit potential scale economies—as can apparently formal institutions that are applied corruptly (see enforcement mechanisms). Formal institutions (such as commercial law at the national or state level) based on competent application of contractarian philosophy and applied equitably across families and across social/ethnic communities can facilitate the growth of "impersonal" transactions and thus can greatly expand the scope for contracting and enhance the range and number of feasible transactions.

> **Source:** Quoted section from Livelihoods Connect—One of a family of knowledge services from IDS. http://www.livelihoods.org/pip/pip/niegloss.html. Remainder provided by the authors.

"**Institutional arrangements** are defined as the forms of contract or arrangement that are set up for particular transactions or holding of assets."

> **Source:** Livelihoods Connect—One of a family of knowledge services from IDS. http://www.livelihoods.org/pip/pip/niegloss.html.

"**Institutional environment** is defined as the broader set of institutions (or 'rules of the game') within which people and organizations develop and implement specific institutional arrangements."

> **Source:** Livelihoods Connect—One of a family of knowledge services from IDS. http://www.livelihoods.org/pip/pip/niegloss.html.

Market versus hierarchy: Why do firms perform some transformation activities in house (that is, use "hierarchy" or "command-and-control" management/contracting systems) while buying or outsourcing (that is, use "market" management/contracting systems) for others? Economists Ronald Coase and Oliver Williamson are closely associated with the proposition that relative transaction costs of using market versus hierarchy in transformation provide the answer to this question: Businesses will use whatever organizational alternatives provide the lowest relative transaction costs. This and related discussions, following Williamson's reformulation (in the 1970s) of Coase's discussion of market versus hierarchy (from 1937) around transaction costs terminology, has given rise to Transaction Cost Economics and NIE as fields of study and has refocused much of the theoretical and empirical work in the field of Industrial Organization. Within the NIE, contract theory explores the forms and roles of contracts in managing relationships among resources and enterprises (individuals) in the transformation process. Thus, the varying forms of institutional arrangements surrounding application of the ESCO model in the Three Country Energy Efficiency Project studies represent case studies in a law, economics, and finance; and these studies are related to fields of inquiry spawned by Coase's original analysis of the choice of market versus hierarchy.

Source: The authors.

"**New Institutional Economics** (NIE) is an interdisciplinary enterprise combining economics, law, organization theory, political science, sociology and anthropology to understand the institutions of social, political and commercial life. It borrows liberally from various social-science disciplines, but its primary language is economics. Its goal is to explain what institutions are, how they arise, what purposes they serve, how they change and how—if at all—they should be reformed."

Source: International Society for New Institutional Economics. http://www.isnie.org/

Organization: "A group of individuals bound by some common purpose to achieve objectives. Organizations include political bodies (political parties, regulatory agencies), economic bodies (firms, trade unions), social bodies (churches, clubs), and educational bodies (schools, universities).

Note that the term 'institution' refers to the rules of the game, whereas 'organization' refers to players of the game."

Source: Ronald Coase Institute, on-line at http://www.coase.org/nieglossary.htm#Organization, with reference to Douglass North (1990), *Institutions, Institutional Change and Economic Performance*, Cambridge: Cambridge University Press, p. 5.

"**Path dependence** refers to the institutions at one place in time being determined by their evolution from earlier institutions. It is defined by North as "a term used to account for the parallel characteristic of an institutional framework that has shaped downstream institutional choices and in consequence makes it difficult to alter the direction of an economy once it is on a particular institutional path. The reason is that the organizations of an economy and the interest groups they produce are a consequence of the opportunity set provided by the existing institutional framework. The resulting complementarities, economies of scope and network externalities reflect the symbiotic interdependence among the existing rules, the complementary informal constraints, and the interests of members of organizations created as a consequence of the institutional framework. In effect, an institutional matrix creates organizations and interest groups whose welfare depends on that institutional framework."

> **Source:** Livelihoods Connect—One of a family of knowledge services from IDS. http:// www. livelihoods.org/pip/pip/niegloss.html. Definition taken from North, D. (1997) "The Contribution of the New Institutional Economics to an Understanding of the Transition Problem." 1997 UNU/WIDER Annual Lecture. http://www.wider.unu. edu/welcome.htm see under 1997 UNU/WIDER Annual Lecture.

Performance contracts (performance-based contracting) are contracts that specify what is to be achieved rather than how it is to be achieved and that usually include performance guarantees that can be specified in either absolute or relative terms. The transition from "how" to "what" is an important element in the ongoing transition from hierarchical, command-and-control management of transformation to the use of market-based contracting for components of transformation processes. The availability of effective technologies and procedures for monitoring and verification of performance are critical to the creation of formal markets for performance contracts. Energy performance contracts developed into a prominent form of performance contract in North America after the energy crises in the 1970s and 1980s. Performance contracts are increasingly used by government agencies to outsource (privatize) the provision of public services.

> **Source:** The authors.

Performance risk is the risk borne by one party to a performance contract that the other party will not fulfill the performance specification.

Performance risk can include both technical risk and credit risk—that is, the risk that the contractor might not be technically or financially capable of fulfilling the specified level of performance. The presence and degree of performance risk is related to the institutional environment in which the transaction (contract) occurs, including not only the formal and informal rules that nominally govern short-term and long-term contracts, but also the enforcement mechanisms that increase the certainty and equitability of transaction fulfillment. Understanding the formal and informal mechanisms of enforcement is required in effectively designing the institutional arrangements that determine whether a particular transaction progresses as initially intended.

Source: The authors.

"**Transaction** refers to the activities that allow/constrain transformation activities. A transaction occurs when two or more parties enter into a contract in which rights and obligations are exchanged. Transactions range from those where all the rights and obligations under the contract take place at a single instant in time, with no rights or obligations remaining outstanding after the instant in time has passed (this is the purest form of spot market transaction) to those which involve a continuous exchange in which reciprocal rights and obligations are part of a permanent contractual arrangement, with no specific termination point—such as in the case of the long-term relationship between landlords and tenants, between plantation owners and their permanent employees, or between a fisheries management authority and individual fishermen. Transactions also include those activities required to define, implement and enforce a given set of property rights over an asset."

Source: Livelihoods Connect—One of a family of knowledge services from IDS. http://www.livelihoods.org/pip/pip/niegloss.html.

"**Transaction costs** are the costs associated with the transactions that are necessary for transformation activity to occur. These costs can be usefully divided into ex ante costs and ex post costs. Ex ante costs involve the costs associated with gathering information about an asset or the good or service derived from its use, and about the other parties involved and the negotiating of and devising a contract or set of rights in such a way as to maximize the likelihood of the contracting partners meeting their obligations under the contract. Measurement costs are an important part of the ex ante cost of a transaction. These are the costs involved in gathering information about the attribute of a commodity or service that is to be the object of a contract and about the reliability and trustworthiness of other parties to the contract. A fisheries management authority needs to obtain information about the state of the natural resource base, as well as about fishermen, before it enters into a contract with fishermen entailing the allocation of fishing rights and the provision of management services in exchange for the undertaking by fishermen to abide by certain rules and regulations. Without such information, the objectives of fisheries management are unlikely to be realized. Ex post costs are those associated with monitoring the performance of any contract or set of property rights associated with the use or exchange [see monitoring and verification discussion in performance contracting], to ensure that the contracting partner is meeting his/her obligations that are incurred once a contract has been negotiated and agreed, and where a transaction is extended over time, and enforcing contracts where the contracting partner fails to meet the obligations stipulated in the agreement. Enforcement costs [see enforcement mechanisms] are incurred when a party to a contract strays from the obligations laid down in the agreement, and are associated with measures taken by the aggrieved party to rectify matters. The exclusion of parties from using resources where ownership is defined to others is another example of an ex post cost."

Source: Livelihoods Connect—One of a family of knowledge services from IDS. http://www.livelihoods.org/pip/pip/niegloss.html.

INDEX

Boxes, figures, notes, and tables are indicated by "b", "f", "n", and "t", respectively.

A

aggregation, 46. *See also* project aggregators
Ahmedabad Electric Company (AEC), 237–38
Alternative Policy Scenario, IEA, 3
ANEEL, Brazil, 135, 235, 236
appraisals. *See* project appraisals; technical appraisals
auto component industry cluster, 198

B

Bank of India, 201
bank products, existing and new, 12
banking sector, 101, 110
   attracting energy efficiency business, 122–24
   Brazil, 64, 105–6, 143
   China, 105, 106, 128
      development of lending business, 73, 125, 146
      improving relationship with ESCOs, 147
   condition of, 12, 105–7

   cost-effective loan structure, 64
   India, 106–7
   Hungary, 128
   local sectors less developed, 104–5
   selecting targets, 82
   support for launching new products, 123–24
   view on energy efficiency investment, 11–12, 103–4
banks, 109–10, 149. *See also* commercial banks
   active role of head office (India), 200–201
   cluster lending, 194–204
   examples of lending programs led by, 124–25
   financing, 107, 110–11
   minimizing transaction costs, 125
   relationship with project developers in HEECP, 176–77
barriers, 6, 50, 64
   to ESCO development, 231–32
BEE. *See* Bureau of Energy Efficiency
BG India Energy Services Private Limited, 251
BgEEF. *See* Bulgarian Energy Efficiency Fund
BNDES, 143
Brazil. *See also* public benefit wire-charge mechanism

273

banking sector, 64, 105–6, 143
contract enforcement, 57t
electricity savings, 154n
energy consumption, 25, 25f
energy efficiency capacity
    development, 90–91
energy efficiency status review,
    144–45
ESCOs, 133, 135
    contracts with utilities, 239t
    government support of, 136
    lighting improvements, 139
    recommendations for, 143–45
    utilities' efficiency investments,
      238f, 239
British Gas, 250
Bucharest Bank Offered Rate, 188n
build-operate-transfer, 114
build-own-operate, 74–75, 114
buildings sector, 214. *See also* public
    buildings
  existing, 44, 47n
    efficiency improvement, 4–5
    improving energy efficiency, 38–42
    interventions, 37t
    thermal renovations, 205
  institutional, 229, 239
  new, 36–37, 41
    components for efficiency, 43–44
    interventions, 37t
    policy and regulatory tools, 38, 38t
    tools for energy efficiency, 42
  restructuring projects, 47n
  retrofit, 44, 210
Bulgaria, 212
Bulgarian Energy Efficiency Fund
  (BgEEF), 187–88
Bureau of Energy Efficiency (India), 90,
    95, 201

C

Canada
  contract enforcement, 57t
  ESCO activity decline, 231
  ESCO development (case study), 69,
    224, 229–30, 229–34

ESCO industry
    development barriers, 136, 231–32
    lessons learned, 231–34
    public buildings market, 233
    Treasury Board, 232
capacity factor of equipment, 86
capacity, local, 9
  in-house
  technical
Carbon dioxide, 4, 27f
Carbon Finance, 30
cash flow, 163, 165–66, 184f
CEEF. *See* Commercializing Energy
    Efficiency Finance
CEMIG, Brazil, 242n
Ceylon Electricity Board (CEB),
    Sri Lanka, 139, 243–45
chauffage agreements, 56, 70, 74, 114, 251
chauffage model risks, 87
China, 165. *See also* ESCO Loan Guarantee
    Program
  banking sector, 105, 106, 128
  contract enforcement, 57t
  developing lending programs, 73,
    125, 146
  EMCA, 223
  energy conservation projects, 162–63
  energy consumption, 25, 25f
  energy efficiency efforts, 90
  energy savings target, 145–46
  EPCs, 213, 217t
    introduction of, 217–18
    shared savings, 215
  ESCO industry, 80, 127, 132–33,
    146–47
    development, 135, 216–18
    in stages, 222
    growth and energy savings, 167
  ESCOs
    adapting to local conditions, 222–23
    full-service, 213–23, 214f
      pros and cons, 220–21
      repayment, 216
    government support of, 136, 221–22
    lessons learned from promotion
      program, 221–23
    pilots, 217–18
    specialized, 214–15
    types of projects, 218, 219f
  recommendations for, 145–47

China Energy Conservation Project, 163, 217
China National Investment and Guarantee Company (I&G), 72, 111, 163
  ESCO Loan Guarantee Program, 165, 168
  loan guarantees issued, 166
  product expansion, 167–68
cluster approach to reduce transaction costs, 97–98
cluster lending, 98, 194–204, 197*f*
  appraisal process, 201
  early approaches, 195–98
  identifying technology options, 199–200
  lessons learned, 202–3
  objectives, 195
  past experience (demonstration effect), 201–2
  project example, 198–99
  recent initiatives, 200
  types, 196
CO_2 emissions, 4, 27*f*
Coase, Ronald, 62*n*, 67, 268
cogeneration projects, 250, 253, 254, 256*n*
commercial banks, 14, 73, 120–25
  lending to housing sector, 211
Commercializing Energy Efficiency Finance (CEEF), IFC, 175
compact fluorescent lamp (CFL) market, 243–45
concessional financing, 13, 119–20
contract enforcement, 56–57, 57*t*, 58
contract monitoring costs, 272
contract theory, 268–69
contracting, 58, 78*n*. *See also* energy performance contracting
contracting institutions, 55
  weak, 78*n*, 158
contracting, relationship-based, 78*n*, 98
core competencies, 53
cost effectiveness, 28, 29*b*, 39.
cost recovery, 29*b*
cost savings finance, 103–4
cost savings for loan repayment, 40*b*
cost-effectiveness, 28, 29*b*, 39
costs. *See also* transaction costs
country institutional context, 158–59
courts, 78*n*

credibility, 55
customized approaches, 118–19

D

deal structuring, 13–14, 60, 69–70, 158, 159, 160
delivery mechanisms, examples of, 71–77
demand-side management (DSM) programs
  example, shifting to natural gas, 140
  IES, 250
  Sri Lanka (case study), 243–45, 244*f*
  utility incentive issue and CEB, 139
Department for International Development, UK, 218
developing countries
  energy demand, 24, 25, 25*f*, 25*t*
  ESCO experience, 16, 134
  investment delivery mechanism model, 68*b*
Developing Financial Intermediation Mechanisms for Energy Efficiency Projects in Brazil, China and India. *See* Three Country Energy Efficiency Project
development finance institutions (DFIs), 10, 15, 129–31
  IREDA, 189–93
  pros and cons to, 191–93, 192*t*
diagnostic reviews of institutional environment, 118–19
discount rate, 6, 50, 58, 59, 62*n*
documentation, 109
Dongying Shengdong EMC, 70
  Waste Gas Power Projects, 246–49, 247*f*, 249
Dongying Shengdong Machinery Company Ltd., 246
donor assistance, 89, 91

E

economic growth, world, 24
economic uncertainty, 58, 61*n*

electricity conservation, 75
Electricity Generating Authority of
    Thailand, 242n
electricity savings, Brazil, 154n
electricity system, India, 257
Eletrobras, 91
EMC. *See* energy service companies
EMC (ESCO) Association, China
    (EMCA), 128, 163, 165, 223
emissions growth, 4, 27f
end users
    outsourcing, 92–93
    pay commercial terms, 13, 119, 126–27
    self-financing, 107
energy audits, 84, 94–95
    credibility of, 201
energy auditors, 88, 89, 90, 94–5,
    148, 201
Energy Conservation Act, India, 148
Energy Conservation Fund,
    Sri Lanka, 244
energy conservation measures, Ohio, 226
Energy Conservation Promotion Fund,
    Thailand, 242n
energy consumption, 46n
    top 10 consumers, 25
energy cost savings, 29bf, 112
energy demand, 25, 27–28
    and security, 23
    developing countries, 25f
    increase and projection, 3, 24, 26
    world by region, 25t
energy efficiency, 3–4, 23–24, 27–30
    and utilities, 75
    dependence on institutions, 51
    holistic view, 145
    lower priority than growth, 59–60
    shifts, 47n
    through restructuring projects, 41
Energy Efficiency Cluster Lending for
    SMEs, 72, 194–204
energy efficiency investment, 5, 167
    importance of institutional environ-
        ments, 51–56
    spontaneous, 77
    through ESCOs, 73–75
    vs. production investment, 59–60
energy efficiency projects, 5, 52, 150
    common traits of retrofits, 45–46
    cost-effectiveness, 28, 29b
    costs and returns, 19–20

diagnostic reviews before design, 153
ESCO services, 131–32
failures, 19, 152
identification of, 79
    energy audits, 84
    unimplemented, 1, 74
implementation barriers, 6
institutional environment
    requirements, 68b
operational suggestions, 152–54
partnering with financial intermediaries,
    120–21
recommendations, 152–53
restructuring, 47n
    vs. standard energy efficiency,
        39b–40b
retrofit, 41, 50
standard energy efficiency, 41–42
strong returns of, 141
system replacements in buildings, 44
use of existing markets, 63
using existing bank products, 12
energy efficiency reviews, previous
    focus, 35
energy efficiency status reviews,
    144–45, 147
energy efficiency terrain, 4, 36
energy efficient mortgage, 48n
Energy Management Company (EMC).
    See energy service companies
energy management, outsourcing, 93
energy performance contracts (EPCs), 56,
    73, 96, 131
    Canada, 224
    China, 167, 213, 217t
        introducing, 64, 217–18
    documentation availability, 109
    FBI projects, 230
    FREE, 183
    guaranteed savings model, 132, 133f,
        136, 227f
    policy and legal environment, 89
    shared savings, 132, 132f, 136, 215, 227f
    sophistication of, 134
    United States, 224, 226–27
    variations emerging, 215
Energy Policy Act of 2005, 229
energy savings, 4, 47n
    China's goals, 145, 167
    cost-effective, 35
    flow of, 103, 132, 133–34

INDEX 277

India, 154*n*
Lithuania, 208
technical potential,
energy service companies (ESCOs), 8, 15–16, 73–75, 131–38
　as financing source, 108
　as project aggregators, 113
　Brazil, 239
　　contracts with utilities, 239*t*
　business models, choice of, 137
　Canada, 224–34
　capacity development in Hungary, 172
　capital needs, 138
　case studies, 74
　China, 80 (*see also* full-service *below*; ESCO Loan Guarantee Program)
　　adapting to local conditions, 222–23
　　developing commercial lending, 165
　　EPCs, 64, 217*t*
　　investments in lighting, 47*n*–48*n*
　　types of projects, 219*f*
　development, public building programs, 224–25
　EMCA, 163
　fostering relationships with banks, 147
　full-service, 78*n*, 131–32, 213–23, 214*f*
　India Capacitor Leasing, 259
　lessons learned, Canada and United States, 231–34
　partial financing mechanism, 132
　project development, 133–34
　　and technical assistance, 95–96
　project financing, 135
　skills needed, 137
　uncertain future in India, 148
　United States, 224–34
energy service company (ESCO) industry, 132–33, 162
　Brazil, 135, 136, 145
　China, 127, 132–33, 135, 146–47, 167, 216–18, 218 (*see also* ESCO Loan Guarantee Program)
　developed in stages, 222
　government support essential, 136, 221–22
　Hungary, 176
　India, 135–36
　United States
　　initial obstacles, 226
　United States and Canada, 231–32

energy service company (ESCO) industry association, 133, 223
enforcement, 51, 266–67. *See also* contract enforcement
environment, economic, 57, 58, 160, 194
　institutional, 55, 78*n*, 98
environmental effects of energy production, 26
environmental impact, 26, 28, 210
environmental improvement, reasons for energy efficiency, 28
equipment leasing, 75
equipment vendors, 94
equity finance, 108
ESCO Association, China (EMCA), 128, 135, 163, 223
ESCO Loan Guarantee Program, China, 111, 140*n*, 146–47, 162–63, 164*f*, 165–69
　bank involvement, 169
　comparison to HEECP, 127–29
　impact on ESCO industry, 167
　implementation and pros and cons, 168–69
　results of, 166–68
　Special Fund inflows and outflows, 165
　structural overview, 164*f*
escrow accounts, 12, 113
European Bank for Reconstruction and Development (EBRD), 188, 212
evaluation. *See* monitoring and evaluation

F

facilities. *See* buildings
failure, 19, 121*b*–122*b*
Federal Buildings Initiative (FBI), Canada, 230
Federal Energy Management Program (FEMP), U.S., 228–29
finance, function of, 45–46
financial institutions, nonbank, 107
financial intermediaries, assistance to, 126
financial intermediation, 19, 153, 207
financial markets, 58, 61*n*
financial mechanisms. *See* investment delivery mechanisms

financial sector, increasing lending, 178
financiers, 82–83, 113
   outsourcing, 93–94
financing, 7, 10–13. *See also* project financing
   and guarantees, 222
   challenges presented by banking sector, 105–7
   developing and sustaining flows, 110
   Dongying Shengdong EMC Waste Gas Power Projects, 248
   end users pay commercial terms, 13, 119, 126–27
   ESCO Loan Guarantee Program, 167
   ESCOs, 131–32, 135–36
   IFI tools, 150–51
   institutional options for delivering, 107–10
   integrating with project development, 13–14
   project promoters, 102–3, 111
   sources, 101
   with existing products, 110–11
FREE. *See* Romanian Energy Efficiency Fund

G

glass cluster, 199–200
Global Environment Facility (GEF), 18, 222
   ESCO Loan Guarantee Program, China, 163, 164*f*
   Hungary, 127, 128, 170–80
   IREDA grant, 189, 190
   OP 5, 49–50, 63, 151–52
   Romania, 181–83, 184*f*, 185–88, 187*t*
global warming, 26
government, 4, 55
   providing incentives for utilities, 17
   support, 149
      for 3CEE critical, 17–18
      for ESCOs, 136
      for IREDA, 148
      importance of, 142–43
      use of ESCOs, 96
grant financing, 151, 222
greenhouse gases, 26, 27*f*, 28

growth, 24–27
   compromising energy efficiency, 37
   higher priority than efficiency, 59–60
guarantee companies, 220
Guarantee Facility Agreements (GFAs), 127, 170, 171
guarantee products, pros and cons, 178
Guarantee Program Implementation Agreement, 163, 165
guarantees, 8, 14–15, 72, 124, 170. *See also* loan guarantees; performance guarantees

H

hierarchy approach, 58, 66
hierarchy *vs.* market, 268–69
homeowners associations (HOAs), 205, 207, 210–11
Housing and Urban Development Foundation (HUDF), 211–12
human capital, 86, 137
Hungary Energy Efficiency Co-Financing Program (HEECP), 127–29, 170, 212
   evolution of parameters, 173
   institutional arrangements, 170–72, 171*f*, 174–76
   lessons learned, 179–80
   objectives, 178
   phases, 172
   project developers relationship with banks, 176–77
   results, 174*f*
   slow start up, 179
Hungary Energy Efficiency Guarantee Fund, 170–72, 171*f*, 173*t*, 174–80, 174*f*
Hungary Energy Efficiency Loan Guarantee Program, 72
Hungary, ESCO industry, 176

I

I&G. *See* China National Investment and Guarantee Company

identification. *See* project identification
IEA. *See* International Energy Agency
IES. *See* Iqara Energy Services
IFC. *See* International Finance Corporation
IFIs. *See* international finance institutions
implementation, 92, 141, 189
  barriers, 6, 50, 64
  labor issues, 153
implementation plan, 86
in-house capacity, 57, 66, 194
incentives, 69, 117
  for banks, 122–24
  for utilities, 17
  generating deal flow, 13–14
  U.S. agencies and staff, 228
India. *See also* cluster lending
  banking sector, 106–7
  BEE, 90, 95, 201
  cluster lending by banks to SMEs (case study), 194–204
  contract enforcement, 57t
  developing lending businesses, 148–49
  electricity system, 257
  Energy Conservation Act, 148
  energy conservation and management, 90
  energy consumption, 25, 25f
  energy efficiency status review, 147
  energy savings, 154n
  ESCO industry, 135–36, 148
  ESCO models, 133
  recommendations for, 147–49
  scaling up implementation, 189
  SME lending programs, 80, 124–25
  SME sector, 195
  support for participating banks, 149
India Capacitor Leasing, 257–59, 258f
India Renewable Energy Development Agency (IREDA), 72, 81, 129, 189–93, 193n
  government support of, 148
  institutional arrangements, 190f
  pros and cons, 191–93, 192t
  repayment arrangement, 258
  revolving fund, 189
industrial organization, 67
industrial renovation projects, 42–43
industry interventions, 37t
information and research costs, 53, 272
institutional arrangements

ESCO Loan Guarantee Program, 163, 165–66
FREE, 183, 184f, 185–86
Hungary Energy Efficiency Guarantee Fund, 170–72, 171f, 174–76
IES, 251–54, 252f
India Capacitor Leasing, 258f
IREDA, 190f
Lithuania Energy Efficiency and Housing Pilot Project, 205–8, 206f
public benefit wire-charge mechanism, 235–39
institutional capacity, 88–92, 187
institutional context, 158–59
institutional development, 159–60, 142, 161n
institutional environment, 51–56, 70, 159
  customized approaches, 7, 19, 118–19
  effect on incentives, 69
  lack of transparency effects, 78n
  requisites for energy efficiency projects, 68b
institutional frameworks, challenges due to, 56–59
institutional mechanisms, 19–20, 66
  to deliver investments, 117, 150–51
institutional sector, 234n
institutional transplantation, 61n, 118–19
institutions, 60n–61n, 67, 77n
  formal and informal, 267–68
  strength and contract enforcement, 56–57
  weak, 54, 61n
  contracting, 57, 58, 78n
internal rates of return (IRR) for retrofit projects, 46
International Bank for Reconstruction and Development (IBRD), 39b, 151, 189
International Development Association (IDA), 151, 189
International Energy Agency (IEA), 3, 24, 26, 27
International Finance Corporation (IFC), 121, 151
  CEEF, 175
  HEECP, 127, 128, 170, 172
international financial institutions (IFIs), 18, 124, 126, 149–152
interventions, 42–43, 76
  by economic sector, 37t

investment. *See also* energy efficiency
 investment
investment barriers, 50
investment costs, retrofit projects, 45
investment delivery, 7
 model, 65–70, 68*b*
investment delivery mechanisms,
 1, 117, 158
 deal structuring innovations, 60
 development of, 7–8, 68
 institutional development issue, 142
 integrating project development and
  financing, 13–14
 tailoring, 70
 types of, 8, 14–17, 71–72
investment efficiency, 66
investment financing mechanisms, 10
Iqara Energy Services (IES), Brazil,
 74–75, 250–56
 institutional arrangements and results,
  251–54, 252*f*
 lessons learned, 255–56
 projects, 253*t*
IREDA. *See* India Renewable Energy
 Development Agency
IRR. *See* internal rate of return

life-cycle cost, 28–9, 29*b*
lighting, 28, 139, 238
line-of-credit, 121*b*–22*b*, 124, 146
 IREDA, 189–93
 Lithuania, 72
 project, China, 125, 222
Lithuania Energy Efficiency and Housing
 Pilot Project (case study), 72, 81,
 205–12
 follow-up, 209–10
 institutional arrangements,
  205–6, 206*f*
 lessons learned, 210–12
 results, 207–9, 208*t*
load management program, 243
loan financing, 8, 72–73
loan funds, specialized, 129–31
Loan Guarantee Program of the China
 National Investment and
 Guarantee Company, 72
loan guarantees, 111, 126–29
loan structuring, 40*b*, 220
loans, 64, 126–27. *See also* repayment

L

labor intensity, 19, 129, 153
lamps, sale of CFLs, 243–44
Lanka Electricity Company, 244
Lawrence Berkeley National Laboratory
 survey, 229
leasing, 75, 180*n*, 228
 companies, 107–8
lending, 120, 124–25. *See also* cluster
 lending
 business development in Hungary, 170
 commercial banks, 14
 commercial sector increase, 178
 developing in China, 125, 146, 165
 IFIs, 150–51
 in India, 148–49
 incentives for banks, 122–23
 one-stop shops, 129–30
 to HOAs, 210–11
 to SMEs, 200–202

M

market barriers, 49–50
market contracting. *See* outsourcing
market development, 9, 96–97, 175–76
market entry, benefit of first entrant, 99*n*
market expansion, IREDA, 189
market model, 53
market opportunities to bankable
 solutions, 177–78
market outreach, 82–84
market selection, 80–82
market sustainability, 113, 119
market *vs.* hierarchy, 62*n*, 268–69
markets, 63, 110
 creating, 6, 83
 and broadening, 53–54
 governing of, 51–52
 missing and incomplete, 57–58, 62*n*
 production *vs.* productive efficiency, 59
measurement costs, 272
model, investment delivery, 65–70, 68*b*
Modern Portfolio Theory, 61*n*
monitoring and evaluation, 76, 272

multifamily buildings. *See* residential buildings
multilateral development banks (MDBs), 108

N

natural gas use, shifting to, 140
natural gas, attracting customers, 250
Natural Resources Canada, 230, 232
New Institutional Economics (NIE), 65, 77n, 268, 269
nonbank financial institutions (NBFIs), 107

O

Ohio, U.S., 226
Ontario Hydro, 229–30
Operational Program 5 (OP 5, GEF), 49–50, 63, 151
outsourcing, 58, 66, 268–69
  options, 92–96

P

packaging agents. *See* energy service companies
partial guarantees, 8, 14–15, 72, 124, 170
partnerships, strategic, 98
path dependence, 67, 269–70
peak load requirements, reducing, 139
performance contracts, 228–29, 270. *See also* energy performance contracts
performance guarantees, 113, 114n, 257–59
performance risk, 270–71
pipeline development, 186, 192
policy, public 39, 44
  design for energy industries, 42–43
  EPCs, 89
  restructuring projects, 40b
  tools, 38, 38t, 42

political uncertainty, 58, 61n
power factor improvements, 257, 259
price variation risk, 87
principal-agent problem, 47n
PROCEL, Brazil, 91, 237
production investment *vs.* efficiency investment, 59–60
program designers, 68
project aggregators, 113, 132, 250–51. *See also* project packaging
project appraisals, 39b, 40b, 95
project design, 158, 187
  and risk allocation, 87
  and technical appraisals, 8–10
  data provided by energy audits, 84
  flexibility, 153
  identification, xi, 45, 46, 194, 228
  partnering with financial institutions, 120–21
  risks to consider, 85–88
  to reduce transaction costs, 97
project developers, 6, 176–77. *See also* project promoters
project development, 9, 10, 79
  and implementation, 73–75
  communication between financiers and technical experts, 82–83
  costs, 96–97
  deal structuring, 70
  ESCOs, 133–34
  institutional capacity, 88–92
  integrating with financing, 13–14, 92
  outsourcing, 94–96
project failure example, 121b–122b
project financing, 10–13, 74
  pros and cons of ESCOs, 16
  through cost savings, 103–4
project implementation plan, 86
project opportunities, capturing, 5–6
project packaging, 150. *See also* project aggregators
project promoters, 102–3, 111. *See also* project developers
project size, 52
Project Uptech, 111, 194, 196, 202
projects. *See* energy efficiency projects
property rights, 61n
PROESCO, 94, 143–44
public benefit wire-charge mechanism (case study), 235–42

allocation of revenues,
 235–36, 236t
institutional arrangements, 235
pros and cons, 239–40
reform suggestions, 241b
public buildings sector, 224, 233
retrofits, 229–30
public goods approach to reduce
 transaction costs, 98, 99n
public lighting, 238, 242n

R

Raiffeisen Leasing, 180n
Reference Scenario, IEA, 24, 26
regulatory tools, 38, 38t, 42
relationship-based contracting, 78n, 98
renovation projects. *See* retrofit projects
repayment, 248–49, 258, 259
  Chinese ESCOs, 216
  issues, 110–14
  mitigating risks, 12–13
  through energy cost savings, 40b
  through utility bills, 243–45
replication, 46, 97
research costs for projects and
 transactions, 53
Reserve Bank of India (Central Bank),
 107, 200
residential buildings, 205, 210
residential energy use, 46n
residential lending, 211
residential system replacement
 projects, 44
restructuring projects, 39b–40b, 41, 47n
Retail Gas Program, 180n
retrofit projects, 41, 42–43, 66
  barriers to implementation, 50
  common traits of, 45–46
  factors for, 65
  United States and Canada, 224
review, diagnostic, 152–53
revolving funds, 15, 129–31
  FREE, 185
  IREDA, 189
risk
  allocation, 87
  credit, 122b, 168–69, 172, 178, 271

financial, 143–45, 165 (*see also*
 guarantees)
management, 123, 159
repayment (*see* repayment)
markets, 57–58
perceived, 97, 126, 192
price variation, 87
project design, 85
reduction, 12–13, 113, 220
technical, 137
Risoe Centre on Energy, Climate and
 Sustainable Development, 30
Romania, 181, 182t
Romanian Energy Efficiency Fund
 (FREE), 72, 81, 181–88
  institutional arrangements, 183, 184f,
   185–86
  lessons learned, 186–87
  pros and cons, 187t
  results, 183t
  revolving debt fund, 185
Russia Enterprise Housing Divestiture
 Project, 212n

S

Saha Sprague Limited (SSL), 257–58
scaling up, 28–30
schools, 225, 226
Second China Energy Conservation
 Project, 162, 218
Second Renewable Energy Project, India,
 189
security, of energy, 23, 26
Shandong EMC, 216
Shengli Oil Machinery Company, 246,
 247
small and medium enterprise (SME)
 lending programs, 80, 124–25, 195
small and medium enterprises (SMEs),
 cluster lending (case study),
 194–204, 203n
Small Industries Development Bank of
 India (SIDBI), 196, 197, 198, 199,
 200
small-scale industry (SSI), 203n
specialization and cluster approach,
 97–98
specialized loan funds, 129–31

Sri Lanka DSM Project, utility bill as
    repayment mechanism, 69,
    243–45, 244f
Sri Lanka, load profile, 245n
SSL. *See* Saha Sprague Limited
State Bank of India (SBI), 111, 202, 204n
    auto cluster at Pune, 198
    energy auditors, 95, 201
    Project Uptech, 196
subsidies, 13, 119, 209
Super Energy Savings Performance
    Contracts (Super ESPCs), 228–29
sustainability, 150, 179

T

TBSE. *See* Technology Bureau of Small
    Enterprises technical appraisal,
    8–10
technical assessments, 91–92, 131–32
    credibility, 98
    for banks, 124
    outsourcing, 94–95
technical assistance (TA), 128, 211, 218
    cluster lending, 199–200
    importance in HEECP, 172, 177–78
    to support banks, 123
technical capacity, 88, 89–91, 91
technical experts, 82–83, 98, 198–99
technical performance, 85
technical skills, organization of, 89–90
Technology Bureau of Small Enterprises
    (TBSEs) 202, 204n
thermal renovation, difficulty of, 210
Three Country Energy Efficiency (3CEE)
    Project, 2, 30, 235
    communication between efficiency
        promoters and banks, 17–18,
        102, 142
    equity finance, 108–9
    importance of government support,
        142–43
    lessons learned, 31
    partnering with financial institutions,
        121
transaction cost economics, 67, 268
transaction costs, 12, 104, 271–72
    cluster lending, 197
    definition of, 53, 54

impacts on industrial organization, 67
issue for banks, 125
minimizing, 96–98
reducing through concessional
    financing, 119–20
Transaction Guarantee Agreements
    (TGAs), 170
transactions, 271
transition, 110, 157, 270
transplantation, 118–19
transportation interventions, 37t
transportation sector, limited scope for
    efficiency investments, 44–45

U

Union Bank of India, 201
United States, 137
    contract enforcement, 57t
    DSM programs, 138–39
    early EPC projects, 226–27
    ESCO development
        agency and staff incentives, 228
        case study, 69, 224–29, 231–34
    ESCO industry, 136
        amount of energy efficiency
            projects, 229
        barriers to, 231–32
        initial obstacles, 226
        guaranteed savings projects, 228
        retrofits of federal facilities, 226
utilities, 75
    Brazil
        contracts with ESCOs, 239t
        efficiency investments by sector, 238f
        minimizing losses, 239
        public benefit wire-charge
            mechanism, 235–37
        reform suggestions, 241b
    CEB, 243–45
    incentive for DSM programs, 139
    incentives for energy efficiency, 17, 240
    utility bill as repayment mechanism, 114,
        243–45
utility demand-side management (DSM)
    programs, 8, 16–17, 75–76, 136
    case study, 243–45, 244f
    pros and cons, 138

V

value added, 47n

W

wire-charge program, 136, 235–37
World Bank, 30, 151, 162
    developing lending programs, 125
    Doing Business, 56, 61, 160
    IREDA support, 189
    Line of credit project, China, 125, 222
    Lithuania project, 205
    Romania Energy Efficiency Fund, 181–88
    Second Energy Conservation Project, China, 162, 218
    thermal renovation project, 212n
World Energy Outlook, 3, **27**

ECO-AUDIT
Environmental Benefits Statement

The World Bank is committed to preserving endangered forests and natural resources. The Office of the Publisher has chosen to print **Financing Energy Efficiency** on recycled paper with 30 percent postconsumer fiber in accordance with the recommended standards for paper usage set by the Green Press Initiative, a nonprofit program supporting publishers in using fiber that is not sourced from endangered forests. For more information, visit www.greenpressinitiative.org.

Saved:
- 11 trees
- 7 million BTUs of total energy
- 932 lbs. of CO_2 equivalents greenhouse gases
- 3,876 gallons of waste water
- 498 lbs. of solid waste